Cambridge Elements

Elements of Paleontology

W0227719

CONFRONTING PRIOR CONCEPTIONS IN PALEONTOLOGY COURSES

Margaret M. Yacobucci
Bowling Green State University, Ohio

Paleontological
S O C I E T Y

CAMBRIDGE
UNIVERSITY PRESS

CAMBRIDGE
UNIVERSITY PRESS

University Printing House, Cambridge CB2 8BS, United Kingdom

One Liberty Plaza, 20th Floor, New York, NY 10006, USA

477 Williamstown Road, Port Melbourne, VIC 3207, Australia

314–321, 3rd Floor, Plot 3, Splendor Forum, Jasola District Centre,
New Delhi – 110025, India

79 Anson Road, #06–04/06, Singapore 079906

Cambridge University Press is part of the University of Cambridge.

It furthers the University's mission by disseminating knowledge in the pursuit of
education, learning, and research at the highest international levels of excellence.

www.cambridge.org
Information on this title: www.cambridge.org/9781108717830
DOI: 10.1017/9781108681391

First published 2018

A catalogue record for this publication is available from the British Library.

ISBN 978-1-108-71783-0 Paperback
ISSN 2517-780X (online)
ISSN 2517-7796 (print)

Confronting Prior Conceptions in Paleontology Courses

Elements of Paleontology

DOI: 10.1017/9781108681391
First published online: October 2018

Margaret M. Yacobucci
Bowling Green State University, Ohio

Abstract: People hold a variety of prior conceptions that impact their learning. Prior conceptions that include erroneous or incomplete understandings represent a significant barrier to durable learning, as they are often difficult to change. While researchers have documented students' prior conceptions in many areas of geoscience, little is known about prior conceptions involving paleontology. Here, data on student prior conceptions from two introductory undergraduate paleontology courses are presented. In addition to more general misunderstandings about the nature of science, many students held incorrect ideas about methods of historical geology, Earth history, ancient life, and evolution. Of special note are student perceptions of the limits of paleontology as scientific inquiry. By intentionally eliciting students' prior conceptions and implementing the pedagogical strategies described in other Elements in this series, we can shape instruction to challenge this negative view of paleontology and improve student learning.

Keywords: Pedagogy, STEM Education, Paleontology, Nature of Science, Evolution, Undergraduate

ISBNs: 9781108717830 (PB), 9781108681391 (OC)
ISSNs: 2517-780X (online), 2517-7796 (print)

Contents

1 The Case of the Naked Clams   1

2 The Challenge of Prior Conceptions   2

3 Exploring Prior Conceptions   5

4 Prior Conceptions in Introductory Paleontology Courses   11

5 Conclusions   23

   References   45

1 The Case of the Naked Clams

As instructors, we often take for granted that our students have an accurate working knowledge of the natural world. Sure, they probably don't know the details of a particular fossil group or significant event in Earth history – that's why they are taking our course! – but they at least have a rudimentary understanding of natural processes, including how scientific inquiry works and the basic biology of plants and animals. Given the inherent time constraints of a semester-long course, we instructors have to assume some background knowledge in order to get to the interesting stuff, right? Unfortunately, this assumption is contradicted by research that shows all students enter a classroom with a variety of prior conceptions about the course topic, many of which are inaccurate. We ignore these prior conceptions at our peril, as the following example illustrates.

I teach an introductory undergraduate course for nonscience majors called "Life Through Time." This lecture-lab course provides an overview of how scientists study the ancient Earth and discusses key events in the history of life on Earth. We talk at length about evolution and natural selection and then explore specific evolutionary events, from the Cambrian Explosion to the origin of tetrapods to the Mesozoic Marine Revolution. Since I study the evolution of marine mollusks, I use molluscan examples frequently in the course. By the time we arrive at the Mesozoic Marine Revolution, about two-thirds of the way through the semester, I had always assumed my students are comfortable with the idea of paleontologists studying fossil shells to understand molluscan evolution. An off-hand comment made by a student to a graduate teaching assistant (TA), however, showed me how wrong I was.

One semester, in our weekly meeting following the lab activity on the Mesozoic Marine Revolution, I asked the TAs how students had done. Did the array of mollusk shells that students studied in the lab effectively show how mollusks evolved various defenses to make their shells more resistant to predation? One TA said that her students had trouble seeing how the shells related to the evolution of clams and snails, since, as one student put it, "they could pick out any shell off the beach." Huh? When asked to clarify, the student explained that clams and snails found their shells by crawling naked up onto the beach and looking at the various shells found there. They picked out one they liked, then crawled back into the ocean. But where did the shells come from? The student patiently explained that shells are rocks (made of the mineral calcium carbonate, just like she had been taught earlier in the semester) that formed by crystallizing out of the seawater when it washes up on beaches. After all, when you go to the beach, that's where all the shells are. The student was confused about how just looking at fossil shells could tell you about the evolution of particular animal groups, since the different shell forms would be randomly distributed among the animals who came to live in them. Other students then chimed in, telling the TA that they had been having the same confusion, but had been afraid to ask about it. In all, about one-sixth of the students in that lab section admitted to this understanding of shells.

How did these students come to think that shells are rocks, not biologically produced skeletons, part of the anatomy of the animal? For starters, common-sense observation. Where does one find shells in nature? On the beach, of course! And every child learns about hermit crabs, who shop around for a suitable shell to call home. So the naive idea that other shelly animals, like clams and snails, similarly find their shells is not unreasonable. The students then brought this prior conception into my course with them. When I said things that contradicted it, such as explaining that shells are biogenic, produced by organisms to be their skeletons (which I did early in the semester), the students ignored or forgot that information because it did not align with their own conception of what shells are. And all my arguments on how paleontologists use evidence from fossil shells to document evolution – a central learning outcome for this course – were then quite ineffective in the face of this fundamental confusion about "naked clams."

This example illustrates how important it is for instructors to identify and explicitly address the prior conceptions that our students hold. In this Element, I focus on the former problem. I review the scholarly research on prior conceptions, discuss methods for identifying them, and present data on common misconceptions students bring to introductory paleontology courses. A wide range of pedagogical approaches that can help to address these misconceptions is presented in the other Elements in this series.

2 The Challenge of Prior Conceptions

Science educators and education researchers have long worked to understand what strategies can help or hinder student learning. Most modern science education approaches are rooted in the ideals of constructivism, first formalized by psychologist Jean Piaget (Piaget, 1967, 1973; Piaget and Inhelder, 1969) and now used as an essential framework for understanding the learning process (Bransford et al., 2000; Weimer, 2002; Donovan and Bransford, 2005; Wiggins and McTighe, 2005; Kuh, 2008; Nathan and Alibali, 2010; Yacobucci, 2012; Dahl, 2018, this series). Constructivism argues that people learn by integrating new information into the framework defined by their existing knowledge to construct their own personal updated conceptual framework. Therefore, learning can only take place when the learner modifies preexisting conceptions to accommodate the new knowledge. This notion led to the development of a conceptual change model by Posner et al. (1982), which was further elaborated by Hewson (Hewson, 1981, 1992; Hewson and Hewson, 1988). Other key papers on conceptual change include those by Driver and Erickson (1983), Driver et al. (1985, 1994), Treagust (1986), Chi et al. (1994), Dole and Sinatra (1998), Duit and Treagust (2003), and Stepans (2008). Treagust and Duit (2008) provide an accessible review of the history of the conceptual change model in science education and empirical evidence for its effectiveness.

The conceptual change model argues that students must first recognize their own prior conceptions and compare them to the concept being taught. If they align, learning takes place relatively easily. But if the new idea contradicts the prior conception, learning is more difficult. As Hewson (1992) put it:

> Learners use their existing knowledge (i.e. their conceptual ecology), to determine whether ... a new conception is intelligible (knowing what it means), plausible (believing it to be true), and fruitful (finding it useful). If the new conception is all three, learning proceeds without difficulty ... If, however, the new conception conflicts with existing conceptions, then it cannot become plausible or fruitful until the learner becomes dissatisfied with the old conceptions. In that event, learning requires that existing conceptions be restructured or even exchanged for the new. (Hewson, 1992, pp. 8–9, emphasis in original)

Following the conceptual change model, then, effective classroom practices will provide students with opportunities to self-reflect on their prior conceptions and evaluate whether those conceptions align with the new information to be learned.

All students enter the science classroom with a host of prior conceptions. These may, in fact, be correct and align well with current scientific understandings, or they may represent incorrect ideas of one sort or another. The research literature on student conceptions uses various terms: 1) prior conceptions; 2) preconceptions; 3) alternative conceptions; 4) naive conceptions; 5) intuitive conceptions; or 6) misconceptions (Cheek, 2010; Francek, 2013; Baldwin and Cooper, 2014). As Cheek (2010) noted, it is important not to assume that any prior conception held by a student is likely to be invalid. Hence, I here use "prior conception" as the general term, reserving "misconception" for demonstrably incorrect ideas. Students come to hold these prior conceptions from a variety of sources, including teachers and instructional materials, but also family members, friends, and various media (Cheek, 2010; Baldwin and Cooper, 2014). Culture and language can also act as important influences on students' conceptual understanding, which has implications for science instruction in diverse classroom settings (Lee, 2001; Solano-Flores and Nelson-Barber, 2001; Luykx et al., 2008; Lee et al., 2009).

Prior conceptions are often deeply rooted and difficult to change, even with instruction (Driver and Easley, 1978; Vosniadou and Brewer, 1992; Chi et al., 1994; Bransford et al., 2000; Donovan and Bransford, 2005). For example, Anderson and Libarkin (2016) found that 22 of 73 questions on the Geoscience Concept Inventory (see later in this Element) showed very small post-instruction gains in a large, national sample. Confronting prior conceptions in the geosciences has some special challenges. Because of its synthetic nature, errors in thinking about geoscience processes may derive from incorrect prior conceptions in other disciplines, like physics, chemistry, or biology (Anderson and Libarkin, 2016). Also, students, even at the undergraduate level, have difficulty thinking in terms of processes and systems. Rather, they tend to focus on learning terminology – they can *name* processes like subduction but cannot explain how those processes actually work (Raia, 2005; Libarkin and Kurdziel, 2006). Paleontology and historical geology courses often use an integrative, Earth system science approach to presenting course material. This systems focus may make deep learning in these types of courses particularly difficult for novice students.

Libarkin (2006) noted that in developing the Geoscience Concept Inventory (GCI), a set of questions used to assess students' prior conceptions in the geosciences, some reviewers thought the questions were much too simplistic for undergraduates. However, many undergraduates have indeed been shown to have exceptionally naive views about how the planet works. Students think volcanic eruptions can only occur in warm climates, clouds are empty vessels that fill up with water or pollution, and Earth's magnetic field is what holds continents and people onto the planet's surface (Libarkin, 2006). Lest one think these misconceptions reflect poor scientific instruction in the United States, Felzmann (2017) found that elite German high school students believed that glacial ice forms when temperatures become very cold and snow "freezes." Intuitive or commonsense concepts (e.g., ice forms from freezing something, magnets pull things together) can lead to incorrect conclusions about natural processes. These ideas are generally invisible to instructors but can have a profound impact on the ability of our students to learn. It was only after years of teaching about the Mesozoic Marine Revolution that I learned about the "naked clam" conception, and only then because a graduate TA thought to probe a student's thinking in the lab and then share that thinking during our weekly instructors' meeting. What else are we missing about how our students think about the history of life on Earth?

3 Exploring Prior Conceptions

Most paleontologists to whom I have told the "naked clam" story have been shocked that anyone could think something so obviously incorrect. Most nonscientists have nodded and said, "yes, I can see where that idea is coming from." Therein lies the problem – we as experts are so far removed from the prior conceptions most of our students hold that it is hard for us to even imagine them (Libarkin, 2006). We just don't think like a novice does. To identify students' prior conceptions, then, we must deploy techniques that make use of students' own reports on their thinking.

3.1 Verbal Explanations

Perhaps the most obvious strategy for determining what prior conceptions students may have is to simply ask them. One might, for instance, make a habit of prompting students to explain the reasoning behind the questions they ask during class. However, students are reluctant to sound "stupid" in front of their instructor and peers, so they are less likely to ask questions rooted in their confusion about basic principles or processes during class time. They are more likely to open up, with some prodding, when talking with an instructor individually during office hours. It can be helpful, then, to invite students to office hours and conduct brief one-on-one interviews in which you probe their understanding of a concept. These interviews can provide the instructor with useful information on, for example, an exam question with which many students struggled, as well as helping the individual students to work through their reasoning.

Still, instructors are usually far removed from the novice-level learner's mind-set, and so may not be able to effectively question students to elicit their prior conceptions. It is more effective to deploy peer undergraduate or graduate student TAs for this task, who can then report their findings back to the instructor. In the "naked clam" example, the smaller-group lab section with a graduate student instructor was a classroom climate in which students were more likely to share their ideas (though even here, it took one student brave enough to broach the topic before other students revealed their similar thinking). The graduate student was able to understand where the student was coming from and ask questions that further elucidated the prior conception. Undergraduate student peer facilitators or learning assistants can also be invaluable informants about student conceptions. During in-class activities, undergraduate assistants can be tasked with circulating among student groups and asking students to explain their thought processes. Since these undergraduate assistants are closer in their knowledge progression to the "novice" students in the course than to the "expert" instructor, they are more likely to think of possible prior conceptions their peers may hold (ideas they may have only recently held themselves) and to be able to draw out student thinking in a nonjudgmental way. Taking this approach a step further, students in a course can be asked to discuss the conceptual basis of an idea or problem with each other, then report on a group's ideas without identifying individual students who held those ideas. The peer instruction movement pioneered by physicist Eric Mazur (Mazur, 1997; Zull, 2004) leverages this ability of fellow students to best understand the thinking processes of their peers; we should make use of this resource!

3.2 Written Explanations

Undergraduate students are generally very good at surface learning, that is, memorizing information in order to parrot it back on assignments and exams without really understanding the reasoning behind the answer (National Survey of Student Engagement, 2005; Nelson-Laird et al., 2008). To elicit student thinking and potential prior conceptions, then, it is useful to ask students to explain in writing *why* they gave the answer they did to a short-answer question and to include on exams and lab activities questions that require a longer written response that asks students to explain their reasoning. In addition to providing data on students' thought processes, requiring students to "write out loud" – work through their ideas as they write a response – can help them identify their prior conceptions and where these conceptions may be leading them to an incorrect understanding (Fulwiler, 1987; McDermott, 2010). Interesting prior conceptions can also be collected as "minute papers" at the end of a class session, by giving students a minute to write down anonymously what ideas they have about the topic or concepts they are confused about. Concept sketches are another effective way of quickly assessing how students are thinking about a problem or process (Johnson and Reynolds, 2005). For a more general sense of what ideas students bring into the classroom, one might create a set of open-ended questions derived from key concepts identified in scientific literacy documents (Climate Literacy Network, 2009; Earth Science Literacy Initiative, 2010). Students can then write about these concepts from their own perspective, perhaps followed up by interviews to further elicit student thinking (Libarkin and Kurdziel, 2001, 2002; Baldwin and Cooper, 2014).

3.3 Formative Assessment Probes and Surveys

Techniques like minute papers are types of formative assessment, activities that instructors use during a learning interval to get feedback on student thinking. This feedback is then used to adjust instruction in ways that improve student learning. If an instructor has some sense of the likely prior conceptions students may hold, formative assessment instruments can be created to determine whether students actually do hold those conceptions.

Formative assessment "probes" are narrowly targeted instruments used to elicit student thinking on one or a few central concepts related to the topic being taught. The use of formative assessment probes in K-12 science education has a lengthy history. Page Keeley and colleagues have developed a large library of simple probes that target particular concepts, mostly under the series name "Uncovering Student Ideas in Science" (Keeley, 2005, 2008, 2015a, 2015b; Keeley and Tugel, 2009; Keeley et al., 2005, 2007, 2008) and on the "Uncovering Student Ideas" website (Keeley, 2011). Figure 1 provides one paleontological example of a formative assessment probe, discussed in Keeley (2015a). Concept probes like this one are designed to be deployed, completed,

Mountaintop Fossil

The Esposito family went hiking on a tall mountain. Mrs. Esposito picked up a shell fossil on the top of the mountain. The fossil was once a shelled organism that lived in the ocean. The family had different ideas about how the fossil ended up there. This is what they thought:

Mrs. Esposito: A bird picked up the organism and dropped the shell as it flew over the mountain.

Mr. Esposito: Water, ice, or wind eventually carried the fossil to the top of the mountain.

Rosa: A mountain formed in an area that was once covered by ocean.

Sofia: The fossil flowed out of a volcano that rose up from the ocean floor.

Whose idea do you most agree with and why? Describe your ideas about how a fossil could end up on the top of a tall mountain.

Figure 1 Sample formative assessment probe (Keeley, 2015a)

and scored quickly, so that the instructor can immediately see how many and which students have particular misconceptions about the topic. In the example in Figure 1, students are presented with four alternative explanations for the occurrence of a fossil shell on top of a mountain. Students must state which of the four explanations they agree with and why. While this probe is simple on its face, it forces students to confront fundamental questions about the nature of our planet. Has Earth's surface always been the same or has it changed over time? How do water, wind, and ice move things around the Earth's surface? How do mountains form? How do fossils form? A quick scan of the probe results can tell an instructor which big-picture concepts need to be addressed in class. A word of caution, though: it has been my experience that fellow faculty experts often get the Keeley probes wrong because they overthink the problem. For example, the explanation given by Mrs. Esposito – that a bird dropped the shell as it flew over the mountain – might be seen as perfectly possible, if unlikely, and therefore cannot be rejected with the evidence provided. (This view ignores the information that the bird picked up the "organism," presumably alive at the time and not a fossil.) In these probes, however, the most *likely* explanation is the "correct" one, in this case, Rosa's explanation that the rocks making the mountain were once under the ocean.

Another formative assessment technique built on the instructor's knowledge of likely student misconceptions is the use of student surveys of prior knowledge. In these surveys, students are presented with a set of statements, which may be accurate descriptions of a concept or common misconceptions, and asked to agree or disagree with them. This surveying technique is also meant to be quick to deploy and score, so that busy instructors can efficiently determine whether most of their students understand a concept or hold one or more particular misconceptions on the topic. Surveys can be used before and after instruction on a topic to determine whether the instruction led students away from misconceptions and toward correct understandings or had no (or even a negative) impact on student understanding.

Note that formative assessment techniques can only improve student learning if they are used intentionally and the results form the basis for instruction. Yin et al. (2008) found that formative assessment instruments were effective for eliciting middle school students' prior conceptions, but teachers often found it difficult to provide meaningful feedback to students or to explain to students *why* their misconception was incorrect. Other teachers in Yin et al.'s (2008) study did not modify their instruction at all based on the results of the assessment. These sorts of problems are likely to be exacerbated in university classrooms, where instructors are not well trained in pedagogical techniques or rewarded for committing time to revising course content.

3.4 Conceptests

Conceptests are short, multiple-choice instruments with questions that each target a specific concept and are designed to determine whether students have a correct understanding of the concept or hold one or more incorrect views (Lindell et al., 2007; Undersander et al., 2017). Conceptests are often drawn from an established concept inventory, a pool of at least 20 questions targeting prior conceptions, but they can also be derived from an instructor's own observations and data. The development and use of concept inventories was pioneered by the physics education community (Hestenes et al., 1992; Lindell et al., 2007); there are now established concept inventories for chemistry (Mulford and Robinson, 2002; Pavelich et al., 2004), biology (Garvin-Doxas and Klymkowsky, 2008; Smith et al., 2008; Smith and Tanner, 2010; Perez et al., 2013), astronomy (Bilici et al., 2011), and oceanography (Arthurs et al., 2015), among others (Libarkin, 2008). The GCI is a validated and reliable set of nearly 200 multiple-choice questions that can be used to assess students' prior conceptions on a variety of geoscience topics (Libarkin and Anderson, 2005, 2007a, 2007b; Libarkin et al., 2005, 2011; Libarkin, 2008; Ward et al., 2010). The GCI is available online, including access to all questions and with opportunities to submit new GCI questions (Geoscience Concept Inventory Wiki, no date).

Writing effective multiple-choice conceptest questions that provide meaningful feedback on student learning takes practice. Because undergraduates are often adept at memorization, surface learning, and test-taking, they may select the correct answer on a multiple-choice question by a process of elimination rather than an understanding of the question. It is essential that the distractor choices be plausible misconceptions (avoiding extremes like "always" and "never"), and formatted in a similar way to the correct answer so students cannot automatically eliminate them as choices (Libarkin, 2008; Anderson and Libarkin, 2016). Three to five options for answers are ideal. Avoid "none of the above" as a correct answer, as it does not reveal anything about whether the students know the actual answer to the question, as well as "all of the above," as it only requires the student to identify two correct options. Use caution with wording of both the question stem and distractors, avoiding scientific jargon and complex sentence construction, so as not to make the question a test of language ability rather than scientific understanding. Incorrect ideas observed in previous students' work in the same course generally make the most plausible distractors on conceptests. Also, conceptest questions that go beyond memorization by requiring students to apply their knowledge to a new problem or context will more effectively reveal student misconceptions. It is important to collect data on student responses over multiple administrations of a question. Distractor options that are consistently ignored by students should be replaced with more plausible ones.

3.5 Published Literature on Prior Conceptions

In addition to assessing one's own students, a variety of published sources can be used to identify common misconceptions in the nature of science, geoscience, and life science. Concept inventories like the GCI described earlier are a good place to start. Several websites also provide lists of misconceptions, including Indiana University's Evolution and the Nature of Science Institutes (ENSI) website (Flammer, 1999), and the excellent websites Understanding Science (Understanding Science, 2017) and Understanding Evolution (Understanding Evolution, 2017) created by the University of California Museum of Paleontology. A large literature reflects a steady stream of research studies that have identified common student misconceptions in the geosciences, including the topics of plate tectonics (Sibley, 2005; Clark et al., 2011; Smith and Bermea, 2012), Earth's interior (Steer et al., 2005; Capps et al., 2013), landscapes and surface processes (Martínez et al., 2012; Sexton, 2012; Jolley et al., 2013), glaciers and ice ages (Felzmann, 2017), geologic time (Trend, 1998, 2000, 2001; Dodick and Orion, 2003; Hidalgo and Otero, 2004; Libarkin et al., 2007; Teed and Slattery, 2011), climate change (Rebich and Gautier, 2005; Lambert et al., 2012; Baldwin and Cooper, 2014; Bodzin et al., 2014; McCuin et al., 2014; McNeal et al., 2014; Reichert et al., 2014), oceanography (Arthurs et al., 2015), and Earth systems (Raia, 2005; Libarkin and Kurdziel, 2006; Sell et al., 2006). For more general overviews of geoscience conceptions research, see Phillips (1991), Schoon (1992), Dove (1998), McConnell et al. (2005, 2006), Petcovic and Ruhf (2008), Reinfried and Schuler (2009), Cheek (2010), Francek (2013), and Wild et al. (2013). Of particular note are the extensive studies on geoscience concepts done by Julie Libarkin and colleagues, including Libarkin and Kurdziel (2001), Dahl et al. (2005), Libarkin (2008), Libarkin et al. (2014), and Anderson and Libarkin (2016). Life science education researchers have also identified common student misconceptions that might be useful to instructors of paleontology and historical geology courses (Anderson et al., 2002; D'Avanzo, 2008; Garvin-Doxas and Klymkowsky, 2008; Smith et al., 2008).

Despite this research base, a notable gap exists in the student prior conceptions literature on topics specific to paleontology, such as the fossilization process, the origin and nature of early life on Earth, major evolutionary transitions, extinction events, and the paleobiology of important fossil groups such as dinosaurs. The data I present in what follows are intended as a first attempt to identify and disseminate common prior conceptions undergraduate students hold about paleontological topics.

4 Prior Conceptions in Introductory Paleontology Courses

Every semester, I teach at least one of two large introductory lecture-lab courses, "Life Through Time" and "The Geologic History of Dinosaurs." These courses have enrollments typically ranging from 65 to 140 students and involve two or three lecture sessions and one two-hour lab session (with 18–24 students) each week. The courses count as lab science courses within Bowling Green State University's (BGSU) general education curriculum. Most students are nonscience majors drawn from across the majors offered at BGSU; typically 10–15% of students are preservice teachers (i.e., education majors). The majority of students in "Life Through Time" are freshmen and sophomores; all class years are well represented in "The Geologic History of Dinosaurs."

I have used a variety of methods to assess student prior conceptions in these courses, including multiple-choice conceptests (Appendix A) and formative assessment surveys asking students to agree or disagree with statements about the nature of science (Table 1) and evolution and natural selection (Table 2, Appendix B). I have also administered an anonymous questionnaire on the last day of class asking students to self-report a misconception they had at the beginning of the semester that the course has corrected (Appendix C). The specific questions, survey items, and data compilation of student responses are provided in the appendices.

4.1 Conceptests

To develop the conceptest questions, I considered the prior conceptions that students stated during class and office hours, and that appeared frequently on lecture and lab worksheets and exam responses. I also consulted the research literature on prior conceptions in the geosciences. I considered best practices for crafting conceptest-style questions, such as: 1) phrase the stem as a question, if possible; 2) use a simply worded stem; 3) use plausible response options and meaningful distractors; 4) keep response option lengths similar; 5) use three to five response options; 6) avoid "all of the above" formats; and 7) use language that is technically accurate but still easily understandable to students (Libarkin, 2008; Anderson and Libarkin, 2016).

The pool of questions was used to create short, five-question quizzes, which were administered in both the "Life Through Time" and "The Geologic History of Dinosaurs" courses four to six times during a semester. The quiz questions covered material discussed in the two to five previous class sessions; hence, these are post-instruction assessments. Quizzes counted for a small portion of the overall course grade. Note that I did not deploy quizzes immediately before midterms or final exams, so some essential topics in paleontology and historical geology, such as the global climate change, the Permo-Triassic extinction, and human evolution, are not covered by these conceptests.

After administering, collecting, and scoring the quizzes, I tallied up the number of students who selected each response option and calculated the percentage of students per response. These percentages were useful for identifying common misunderstandings and helped guide my discussion of the quiz during the next class meeting. Quiz results were then filed so they could be compared to those from courses in subsequent semesters. For this Element, I compiled and pooled student data collected from eight semesters of "Life Through Time" and seven semesters of "The Geologic History of Dinosaurs" (from fall 2011 to spring 2017). Questions (with sample sizes and percentages) are listed in Appendix A, with letter and number coding to uniquely identify each question. For each question, I treated responses with student percentages at or above 10% (highlighted in *italic* in Appendix A) as indicating prior conceptions still held by a nontrivial number of students even after instruction and therefore likely to be both durable and potential barriers to student learning. A few of these misconceptions are highlighted here.

Students frequently think that science is restricted to making observations and compiling facts about the natural world (questions NS1, NS2). These students do not understand science to be a process for developing and testing explanations about the natural world, but perceive science as a static body of facts and terms; their job as science students is to memorize these facts. Related to this conception is the idea that a scientific theory is a potential fact that scientists are not yet sure about (question NS3). A theory may one day be confirmed as a fact but is currently uncertain. It takes intentional instruction to show students that scientific theories are well-supported *explanations*, the true goal of scientific inquiry.

For many students, the ancient past was a time unlike our own (question S1), when terrible catastrophes rocked the planet (question T4). On the other hand, the common uniformitarian adage "the present is the key to the past" gets twisted into "the past is the key to the future" (question F1) – by studying ancient Earth, we can predict what will happen in the future. This is certainly a valid statement, but it is not what is meant by uniformitarianism. Students also think that, since society is (rightfully) concerned about anthropogenic warming today, the present must be an especially warm time in Earth history (question CZ6).

Misconceptions about evolution and natural selection appear to be pervasive. Students equate evolution with the notion of predictable progress, defining evolution as the continual improvement of life on Earth (question EV1), with the expectation that fossils will always show life evolving into more advanced forms over time (questions T5, PZ2). Evolution inevitably occurs if the environment changes, either because the environment itself forces the change (question EV4) or because natural selection causes individuals to acquire traits adapted to those changes so they can survive (question EV5). On the other hand, students also equate natural selection with the "strong" beating out the "weak" for survival or consider evolution by natural selection to mean it is "random" who survives (question EV5). These were all popular ideas even after several classes and a lab exercise on evolution and natural selection designed to expressly target these misconceptions. Clearly, the ideas about evolution that students bring into the classroom are particularly resistant to change. They also impact students' thinking about specific evolutionary events, like identifying the origin of eukaryotic cells as either directly caused by the environment or by the "choice" of organisms to evolve (question PC6).

Students frequently think that Earth is less than 4 billion years old (question PC1) and that Earth's age is determined by dating early fossils (question PC2). They conflate the gases of the solar nebula with the gases of Earth's early atmosphere (question PC3) and the actual source of the early atmosphere (volcanic outgassing) with the source of Earth's oxygen (question PC4). Perhaps students have trouble understanding that atmospheric oxygen is largely derived from bacterial photosynthesis because they think only land plants can photosynthesize (question PZ6).

Accurately understanding the timing of important events in Earth history is difficult for many students, who tend to conflate events into a single time period, mix up the order of events, or assume a nonoverlapping sequence of events rather than a more complex, overlapping pattern. For example, some students think land animals evolved at the same time as the very first animals (question PZ2), claim that dinosaurs preceded mammals, who only evolved as dinosaurs died out (questions MZ4, MZ11), and have difficulty separating tectonic events in time (questions CZ3, CZ4, CZ5).

It is important to acknowledge that some concepts seem to be well understood by undergraduate students. The idea of stratigraphic superposition gave few students trouble (question T1), and most had a decent working knowledge of the basic processes of fossilization (questions F2, F3). Students were also generally able to identify a statement involving supernatural processes as not falsifiable while statements about ancient Earth are, even though ancient Earth is not directly observable today (question NS5). Assessing students' prior conceptions can help identify topics like these that will not need as much class time and attention as more problematic concepts will.

4.2 Nature of Science Survey

In addition to these conceptests, I have also assessed students' prior conceptions through the use of student surveys. The nature of science survey (Table 1) is eye-opening, as it shows widespread misconceptions about the practice of science. The survey consists of 10 statements, taken from Flammer's (1999) science knowledge survey. On the first day of "The Geologic History of Dinosaurs," prior to any instruction on the topic, students were asked to complete the survey by indicating whether they thought each item was true or not true. Students then compared their answers to those of a neighboring student, discussed any disagreements, and recorded their final answer next to their original answer. This pairing approach allowed students to talk through their reasoning with a peer and then change their answer if they wished. The survey was administered in five separate iterations of the course (spring 2013 to spring 2017) to a total of 455 students. The survey items and pooled results are presented in Table 1; items for which at least 10% of students gave the incorrect response are in *italic*. Note that these represent 7 of the 10 items. These items clearly represent common misconceptions about science that people hold.

 As seen with the conceptests, many students (34%) viewed science as a descriptive field focused on collecting facts. Indeed, more than half (52%) of the students agreed with the idea that any study "done carefully and based on observation" can be considered "scientific." The majority of students (63%) considered scientific hypotheses to be merely "educated guesses." Students struggled with the idea of the inherently tentative nature of scientific conclusions (41%) and the notion that scientific inquiry is shaped by the characteristics and experiences of the scientist (33%); they expected science to be focused on the entirely objective determination of "truth." Interestingly, though, very few students (3%) agreed with the statement that once something is "proven scientifically," it can no longer be changed. This internal disconnect around the "truth" of science is worth further exploration in the future.

Table 1 Science knowledge survey items and results pooled across five semesters and 455 students. Items for which at least 10% of students gave the incorrect response are in *italic*. Survey items modified from Flammer (1999).

% of students who gave the incorrect response	Correct response	Item
3.1%	not true	1. Something that is "proven scientifically" is considered by scientists as being a fact, and therefore no longer subject to change.
41.1%	true	*2. Science always provides tentative (temporary) answers to questions.*
52.2%	not true	*3. Any study done carefully and based on observation is scientific.*
62.9%	not true	*4. A "hypothesis" is just an "educated guess" about anything.*
3.5%	true	5. Scientists can believe in God or a supernatural being and still do good science.
1.8%	true	6. Science can study things and events from millions of years ago.
34.1%	not true	*7. Science is most concerned with collecting facts.*
32.5%	true	*8. Science can be influenced by race, gender, nationality, or religion of the scientist.*
11.9%	not true	*9. Disagreement between scientists is one of the weaknesses of science.*
13.6%	true	*10. Scientists often try to disprove their own ideas.*

4.3 Evolution and Natural Selection Survey

While a single pre-instruction survey like the science knowledge survey can reveal a lot about what conceptions students bring into a course, it can be even more valuable to use a paired survey approach, where a survey is administered prior to instruction and then again after instruction. By comparing responses, the instructor can identify which conceptions changed during the course (presumably due to student learning) and which conceptions remained resistant to change. I used this approach to explore the impact of instruction on student conceptions of evolution and natural selection in the "Life Through Time" course. I selected 14 common misconceptions about evolution from the compilation provided on the University of California Museum of Paleontology's Understanding Evolution website (Understanding Evolution, 2017). Students were asked to individually and anonymously complete the survey before the class started discussing evolution, about one-third of the way through the semester (the "pre-topic" survey). Students then completed the same survey, again anonymously, in the last week of the semester, after they had learned about evolution, natural selection, and specific evolutionary events in the history of life on Earth (the "post-topic survey"). For each statement, students indicated whether they agreed or disagreed with it using a five-point Likert scale: 1-Strongly Agree, 2-Tend to Agree, 3-Don't Know, 4-Tend to Disagree, 5-Strongly Disagree. Note that, since each survey item was a misconception, ideally students would disagree with every statement (Likert score of 4–5). Students were informed that the survey was not graded and intended to be anonymous, and that they should feel free to react to each statement as honestly as possible.

I administered the pre- and post-topic surveys during six semesters (fall 2012, spring 2013, fall 2013, fall 2014, fall 2015, and fall 2016). The mean Likert score for each item on the pre- and post-topic surveys was calculated from data pooled across all semesters and the difference between the post-topic and pre-topic scores calculated (Table 2). Items that showed a positive shift (that is, toward "disagree," the desired answer) of at least 0.5 on the five-point scale are highlighted in boldface in Table 2. Those items for which the mean score shifted in the negative direction (that is, toward "agree") are highlighted in *italic*. The percentages of students selecting each response were also calculated (Appendix B, Tables B1 and B2).

In the pre-topic survey, students tended to agree with some common misconceptions about evolution, including that evolution is a theory about the origin of life (item 1), that evolution is a climb up a ladder of progress (item 2), that natural selection involves organisms trying to adapt (item 4), and that natural selection gives organisms what they need (item 5) (Table 2). Item 6, "evolution is just a theory," was a little less popular, likely because we had already discussed the meaning of "theory" in science (as an explanation, not something we are unsure about), although the mean score for this item on the pre-topic survey was still 2.98, that is, just on the "agree" side of the scale. Students were less likely to agree with statements about the status of evolution (e.g., gaps in the fossil record disprove evolution [item 8], evolution is not

Table 2 Evolution and natural selection survey results pooled across six semesters. Items showing a positive shift (that is, toward "disagree," the desired answer) of at least 0.5 on the five-point scale are highlighted in boldface, while items shifted in the negative direction (that is, toward "agree") are highlighted in *italic*. See Appendix B for details of survey administration, sample sizes for each item, and response frequencies for pre-topic and post-topic surveys. Survey items modified from Understanding Evolution (2017).

Survey Item	Post-topic mean	Pre-topic mean	Post-minus pre-mean
1. Evolution is a theory about the origin of life.	2.74	2.14	**0.60**
2. Evolution is like a climb up a ladder of progress; organisms are always getting better.	3.67	2.49	**1.18**
3. Evolution means that life changed by chance.	3.47	3.54	*−0.07*
4. Natural selection involves organisms trying to adapt.	2.86	2.11	**0.75**
5. Natural selection gives organisms what they need.	3.62	2.77	**0.85**
6. Evolution is just a theory.	3.55	2.98	**0.57**
7. Most biologists have rejected Darwinism (i.e., no longer really agree with the ideas put forth by Darwin).	3.63	3.54	0.09
8. Gaps in the fossil record disprove evolution.	4.21	3.95	0.25
9. The theory of evolution is flawed, but scientists won't admit it.	3.95	3.62	0.33
10. Evolution is not science because it is not observable or testable.	4.42	4.15	0.26
11. Evolution leads to immoral behavior. If children are taught that they are animals, they will behave like animals.	4.40	4.42	*−0.02*
12. Evolution supports the idea that might makes right and rationalizes the oppression of some people by others.	3.85	3.64	0.20

Table 2 (cont.)

Survey Item	Post-topic mean	Pre-topic mean	Post-minus pre-mean
13. Evolution and religion are incompatible.	3.57	3.38	0.19
14. *Teachers should teach both evolution and creationism and let students decide for themselves.*	2.83	3.00	*−0.17*

science because it is not observable or testable [item 10]) or negative implications of evolution (e.g., evolution leads to immoral behavior [item 11]).

In the "Life Through Time" course, I shape discussion about evolution and the evolutionary history of life on Earth to expressly target items 1 through 6 on the survey. The post-topic survey bears out this focus, as items 1, 2, 4, 5, and 6 all show relatively large positive shifts in mean score, toward the "disagree" side, demonstrating that these student conceptions are indeed changeable. The exception is item 3, evolution means that life changed by chance, which shows a small negative shift (−0.07) in mean score. I suspect this small shift in the wrong direction is due to students' conflation of "evolution" with "the history of life." They perceive events like the end-Cretaceous asteroid impact as chance events that had a big effect on Earth's biota, and take this to mean "evolution" involves chance events.

Two other items showed a shift in the wrong direction. Just 5% of students agreed with the statement that evolution leads to immoral behavior (item 11) on the pre-topic survey, but 9% of students agreed with it after taking the course (Tables B1, B2). It is not clear why this shift happened, although we do discuss scientific racism and eugenics when we cover human evolution late in the course. These negative aspects in the history of evolutionary thought may be the immoral behavior that students mean (see also item 12). Item 14 on the survey, teachers should teach both evolution and creationism and let students decide for themselves, was a polarizing statement (see Table B1) and the only one to show a shift in mean score fully into the "agree" side of the Likert scale (mean score changes from 3.00 to 2.83). The statement can be considered disagreeable for several reasons: science teachers are generally not trained or qualified to teach religious beliefs, evolution and creationism are not two opposing equivalents, there are many different kinds of creationism (not just one), students are not asked to "decide" on any other science topics, and the fact that evolution happens is an observation, not an opinion up for debate. All these arguments were discussed in class, so why does this item show a negative shift of -0.17 from the pre-topic to the post-topic survey? One possibility is that they were using the "Life Through Time" course as their model. I made a point of telling students that I had seriously weighed the pros and cons of even bringing up creationism in the course. Given the limited time in the semester, shouldn't I just focus on teaching students the science of evolution? I explained to the students that I thought it was important for them to learn something about creationism so they would be better prepared as citizens, parents, and teachers to address the issue should it arise later in their lives. Students may have taken this pedagogical approach as advocating for the idea that teachers should teach both evolution and creationism.

Examining the percentages of students selecting each option can also provide insight into students' thinking (Appendix B, Tables B1, B2). Of particular note are items that appear bimodal (with similar percentages of students on the "agree" and "disagree" sides, indicated in boldface in Table B1). These items were especially polarizing concepts in class, and include the idea that evolution is just a theory (item 6) and that teachers should teach both evolution and creationism (item 14), as discussed earlier in this Element. It is also useful to consider those items with a large percentage of students indicating they don't know whether the statement is correct or not. Just two items on the pre-topic survey elicited "don't know" responses for more than one-third of the students (highlighted in *italic* in Table B1), item 7 (most biologists have rejected Darwinism [i.e., no longer really agree with the ideas put forth by Darwin]) and item 12 (evolution supports the idea that might makes right and rationalizes the oppression of some people by others). In the former case, students may have felt they did not know enough about modern biological research to say whether Darwin's ideas are still accepted or not. In the latter case, some students reported verbally to me that they had never heard that idea before and therefore couldn't say whether it was true or not. Others had heard the idea (in the form of eugenics, for instance) but were not sure whether it logically followed from the science of evolution or not. Note finally that the percentage of students marking "don't know" dropped in the post-topic survey for every item except item 14. Overall, instruction in the course did have an impact on many students' conceptions about evolution.

4.4 Student Self-Reported Prior Conceptions

Finally, I asked students in "The Geologic History of Dinosaurs" to self-report misconceptions they held at the beginning of the semester that they felt had been changed because of their learning in the course. Student responses from four semesters (spring semesters in 2014–2017) were reviewed and categorized into themes, which are compiled in Appendix C. Overall, these self-reported misconceptions reinforce what the other data sources showed. Students thought that science was mostly about collecting facts and that hypotheses are just educated guesses. In addition to common confusions about evolution, such as the ideas that organisms intentionally select which traits to evolve and that the environment directly causes evolution to happen, students reported the notion that some groups (like plant-eating vs. meat-eating dinosaurs) are so different from each other that they could not possibly share a common ancestor. They thought that mammals did not evolve until after dinosaurs died out, and believed that all scientists agree that this extinction was caused by an asteroid impact. As other research has also shown (Francek, 2013), students considered carbon dating to be the only kind of radiometric dating and claimed it can be used to date dinosaurs. They understood Pangea to have existed throughout the dinosaurs' time, not breaking apart until after dinosaurs died out.

Other self-reported misconceptions were specifically about dinosaurs. By far the most common misconceptions students cited were that: 1) all extinct animals (e.g., pterosaurs, marine reptiles) are called dinosaurs; 2) there were only a few types of dinosaurs; 3) all dinosaurs lived in the same place at the same time; and 4) all dinosaurs became extinct at the same time. Some students were strongly influenced by the film *Jurassic Park*, believing that all "raptors" were as large and as intelligent as the ones in the film, that *T. rex* could run as fast as a car, and that paleontologists are now close to being able to clone dinosaurs. Interestingly, other students retained a more old-fashioned view of dinosaurs as scaly, dumb, cold-blooded giants living exclusively in tropical forests.

Students' prior conceptions of the nature of paleontology as a field of science were somewhat alarming. They saw paleontology as a limited field of study – they believed that all paleontologists actually do is look for dinosaur fossils and then dig them up. Students reported the perception that fossils are extremely rare and that the only things you can learn from a fossil are what group it belongs to and how old it is. Paleontology is otherwise just guesswork and speculation because of the lack of evidence and direct observations of ancient life. These negative perceptions of paleontology as a scientific discipline represent a major obstacle in engaging students in a robust exploration of our field's core concepts, findings, and contributions to understanding Earth and its biota. Intentional inclusion of real research, via examples, case studies, and activities based on actual scientific studies, is an important strategy to combat these prior conceptions. Such efforts can pay off: one student reported the prior conception that "science is only interesting to scientists," explaining that the course had changed her mind.

4.5 Summary of Prior Conceptions in Paleontology and Related Fields

Based on the data presented earlier in this Element and in the appendices, I provide a summary list of common misconceptions about science, historical geology, paleontology, and evolution. While certainly not exhaustive, the list identifies incorrect views that appear to be particularly common, durable, and damaging to students' ability to learn new concepts in introductory paleontology courses. These misconceptions therefore make appropriate targets for intentional instruction and student exploration.

1) The goal of science is simply to make observations and collect facts about the natural world.
2) Science can only study things we can observe directly.
3) Science is completely objective and never influenced by the personal experiences of scientists.
4) A hypothesis is just an educated guess about anything.
5) Theories are observations that have not yet been confirmed by other scientists.
6) The ancient past was marked by frequent catastrophes that no longer happen today; more generally, processes that occurred in the past are different from those that occur in the present.
7) Uniformitarianism is the idea that the past can be used to predict the future.
8) All radiometric dating is carbon dating, which can date any material of any age.
9) A rock's radiometric date gives the age at which the atoms in the rock first formed.
10) Earth is less than 4 billion years old.
11) Earth's atmosphere formed from the gases present in the solar nebula.
12) Pangea existed from the time Earth first formed, and only broke apart after the dinosaurs became extinct.
13) Earth is experiencing an especially warm climate today compared to the rest of Earth history.
14) The ocean's rocks and biota are the same at all depths; all ocean rocks and fossils formed in coastal environments.
15) Shells are not part of an organism's body but rather are geological objects.
16) Every paleontologist studies all ancient life.
17) Fossils are extremely rare.

18) Paleontology is just guesswork and speculation because of the inability to observe ancient life directly.

19) Early animals were all simple, primitive precursors to modern descendants.

20) Land plants were the first and only organisms to photosynthesize.

21) Animals that are "warm-blooded" or endothermic always give birth to live young, rather than laying eggs.

22) All ancient reptiles or vertebrates are called dinosaurs.

23) All dinosaurs lived together at the same time and became extinct at the same time.

24) There were very few different kinds of dinosaurs, mostly meat-eaters.

25) Dinosaurs were all very large-bodied, scaly, and "cold-blooded" and lived in tropical forests.

26) The depictions of ancient life in the film *Jurassic Park* are scientifically accurate.

27) Birds evolved from pterosaurs, the "flying dinosaurs."

28) Mammals and birds are closely related; feathers are the same structure as fur, but evolved in birds to help them fly.

29) Mammals did not evolve until after dinosaurs became extinct.

30) All scientists agree that an asteroid impact caused the extinction of the dinosaurs.

31) Evolution is just a theory.

32) Evolution is a theory about the origin of life.

33) Evolution is the idea that life has continuously advanced and improved over time; fossils always show predictable, progressive evolution into more advanced forms over time.

34) Some groups of organisms are so different from each other that they could not possibly share a common ancestor; therefore, not all life on Earth is related.

35) The environment is a physical entity that forces organisms to undergo evolutionary change; if the environment changes, evolution will always happen.

36) Individuals will always evolve to survive environmental change because of some undefined "drive"; organisms "choose" or "try" to evolve traits when it makes sense to do so.

37) Individual organisms can evolve during their lifetimes.

38) Natural selection produces the necessary traits in order to give organisms what they need to survive.

39) Natural selection always weeds out the "weak" and permits the "strong" to survive.

40) Natural selection is an entirely random process; because of "random" events like asteroid impacts, evolution means that life changes by chance.

5 Conclusions

How can instructors shape their pedagogical practices in ways that acknowledge and address students' prior conceptions, like those listed earlier in this Element? Once these conceptions have been identified, it is imperative to provide opportunities for students to reflect on them and go through the process of evaluating the alignment of these conceptions to the new ideas to be learned (Posner et al., 1982; Hewson, 1992). The wide range of active learning strategies available to science instructors, including peer instruction, writing- and drawing-to-learn, guided inquiry, and experiential learning, are all effective ways to encourage student conceptual change (Svinicki, 1995; Donovan and Bransford, 2005; Yacobucci, 2012; Dahl, 2018, this series). The other Elements in this series are a great starting place for learning about these pedagogical approaches. Instructors also need to restructure their courses to focus on fewer concepts, the ones that really matter, in order to provide students with the time and space they need to robustly explore them. The classroom climate should promote respect among the instructor and students so students can safely try out ideas that might turn out to be incorrect as they work to construct their own understandings of these concepts.

First, though, we must identify and understand the prior conceptions most likely to hold back our students' learning. I taught "Life Through Time" for years before I ever heard about the "naked clam" problem. Students were successfully able to parrot back what they had learned about the Mesozoic Marine Revolution and predator-prey escalation on exams without having any understanding of the evidence for it. Unless we are intentional about eliciting prior conceptions, we may never even know what our students are missing. Further data on prior conceptions in paleontology and historical geology are sorely needed, including ideas about major evolutionary transitions, extinction events, climate change through deep time, human evolution, and the types of evidence paleontologists use to reconstruct the history of life on Earth, including stable isotope biogeochemistry, stratigraphy, and phylogenetic techniques. By intentionally assessing our students' prior conceptions, we can improve both their learning and our field.

Appendix A
Conceptest Questions for Introductory Paleontology and Historical Geology Courses

The 69 questions listed here are derived from those deployed in two introductory paleontology courses for nonscience majors, "Life Through Time" (pooled over eight semesters from fall 2011 to fall 2016) and "The Geologic History of Dinosaurs" (pooled over seven semesters from spring 2011 to spring 2017). Course enrollments ranged from about 65 to 140 students per semester.

Questions were presented to students as a short, five-question, multiple-choice quiz, given *after* the topics covered by the questions were taught. Quizzes were given four to six times per semester and counted for a small proportion of the students' course grades. Since quizzes were not given on material covered immediately before a midterm or final exam, some essential topics are not represented by these questions (e.g., Earth's climate through time, modern global warming, Permo-Triassic mass extinction, human evolution).

The total number of students who answered a question is listed in parentheses after the question stem. These totals range from 41 to 754 student responses per question (mean = 240, median = 195).

The percentage of students selecting each response option is indicated after the response. Note that these percentages may add up to more than 100% when different response options were used in different iterations of the quiz. Response options for each question are listed in descending order of student percentages (except for "all of the above"–style questions).

The correct answer is in **boldface**. Distractor answers attracting 10% or more of responses are in *italic*. These common distractor answers reflect misconceptions students have about the topic.

Nature of Science 5 questions

NS1. Which of the following statements <u>accurately</u> reflects the scientific process? (N = 508)

    A. Science provides tentative statements only (54.7%)
    B. *Science can only study things we can observe directly (31.2%)*
    C. *Science is mostly about collecting facts (10.6%)*
    D. Science proves things true (8.3%)
    E. Scientific hypotheses are just guesses (7.6%)

NS2. Which of the following is the best description of science? (N = 232)

 A. A process for developing and testing explanations for the natural world (65.5%)

 B. *The careful observation and collection of facts about the natural world (28.4%)*

 C. Any study done carefully and based on observation (3.4%)

 D. A process for proving facts true by doing experiments (2.6%)

NS3. A good definition for the word "theory," as scientists use it, would be: (N = 754)

 A. An explanation for some aspect of the natural world (70.2%)

 B. *An observation that has not yet been confirmed by other scientists (17.6%)*

 C. A controversial hypothesis about why the world looks the way it does (9.4%)

 D. A law of nature, like Newton's law of gravity (2.8%)

NS4. What is absolutely required for an idea to be considered scientific? (N = 274)

 A. It must be falsifiable (85.4%)

 B. *It must be about something we can observe directly (10.9%)*

 C. It must have lots of facts supporting it (1.8%)

 D. It must be quantifiable (1.8%)

NS5. Which of the following statements is **NOT** a falsifiable claim? (N = 163)

 A. Life on Earth arose by the intervention of a divine being (82.2%)

 B. The Grand Canyon formed quickly during a global flood (8.6%)

 C. An asteroid struck Earth 65 million years ago (6.1%)

 D. Humans evolved from an ape-like ancestor 6 to 8 million years ago (3.1%)

Stratigraphy and Geologic Time 11 questions

T1. Steno's principle of superposition states that: (N = 152)

 A. Younger rock layers sit on top of older rock layers (94.1%)

 B. Rocks are originally flat-lying and then are folded by tectonic processes (4.6%)

 C. Older rock layers sit on top of younger rock layers (1.3%)

 D. Rocks are originally curved and then are flattened by tectonic processes (0.0%)

T2. An unconformity represents: (N = 48)

    A. A surface where erosion took place in the past (79.2%)

    B. *A surface where deposition took place in the past (14.6%)*

    C. A surface where volcanic activity took place in the past (4.2%)

    D. A disagreement between two scientists (2.1%)

T3. An unconformity in the rock record represents: (N = 336)

    A. A time of erosion at Earth's surface (75.6%)

    B. *Matching rocks of the same age in two different places (10.4%)*

    C. A catastrophic extinction event (8.6%)

    D. Deposition of sediment under water (5.4%)

T4. As James Hutton argued in 1795, unconformities show that Earth must be very old because: (N = 218)

    A. Unconformities form by erosion, which we observe to be a slow process today (66.1%)

    B. *Unconformities require great catastrophes to form, events which happened long ago (21.1%)*

    C. Unconformities have ancient fossils in them (7.8%)

    D. We know that rocks are old and unconformities are part of the rock record (5.0%)

T5. Faunal succession is the idea that: (N = 521)

    A. Fossil species always appear in the same order within rock layers (67.9%)

    B. *Fossils progressively evolve into more advanced forms over time (15.2%)*

    C. *Extinction is impossible; instead, one species simply evolves into another one (13.1%)*

    D. Fossils are of little use in dating or correlating sedimentary rocks (3.8%)

T6. Fossils can be used to place rocks in a time sequence because: (N = 449)

    A. Some fossil species are only found in a narrow portion of the rock record (61.5%)

    B. *Fossil groups appear within rocks in different sequences, depending on where the rocks are (31.0%)*

C. The same fossil groups often disappear and reappear throughout time (4.9%)

D. Extinction is a rare event (2.7%)

T7. The radiometric age of a rock is: (N = 208)

  A. The date the rock last cooled from a melt (62.9%)

  B. *The date when the atoms in the rock first formed (23.8%)*

  C. The age of the oldest fossil in the rock (7.9%)

  D. The date the rock was deposited in the ocean (1.5%)

T8. In measuring radioactive isotopes, Rock X is found to have a daughter: parent ratio of 2:8 while Rock Y has a ratio of 5:5. Which rock is <u>older</u>? (N = 108)

  A. Rock Y (82.4%)

  B. Rock X (5.8%)

  C. The rocks are the same age (0.9%)

  D. It depends on which type of isotope was measured (0.9%)

T9. A radioactive isotope system has a half-life of 250 million years. Studying a rock sample, you find the ratio of parent to daughter is 25:75. How old is the rock? (N = 48)

  A. 500 million years old (59.3%)

  B. *125 million years old (19.8%)*

  C. *1 billion years old (17.4%)*

  D. 250 million years old (3.6%)

T10. If a mineral crystal originally contained 80 atoms of a radioactive isotope of uranium, after <u>two</u> half-lives, how many atoms of uranium would be left? (N = 52)

  A. 20 (84.6%)

  B. 80 (5.8%)

  C. 0 (5.8%)

  D. 40 (3.8%)

T11. If a mineral crystal originally contained 80 atoms of a radioactive isotope of uranium, after <u>three</u> half-lives, how many atoms of uranium would be left? (N = 110)

  A. 10 (79.1%)

   B. 20 (19.1%)
   C. 5 (1.8%)
   D. 40 (0%)

Sedimentary Rocks 8 questions

S1. Charles Lyell's principle of uniformitarianism (a.k.a. "actualism"):
    (N = 106)

   **A. Notes that present-day geologic processes, given enough time,
        were also important in forming ancient rocks on Earth (71.7%)**
   B. *States that although ancient and modern rocks look very similar, the
        physical processes that created them are not the same (14.2%)*
   C. Implies that Earth's natural laws have varied significantly over the
        past 4.5 billion years (8.5%)
   D. Is no longer used by modern geologists (5.7%)

S2. The grain size of a sedimentary rock is typically a key feature because:
    (N = 151)

   **A. It reflects the environment in which the sediments were depos-
        ited (66.2%)**
   B. *It indicates which specific chemical weathering process produced
        the sediment (17.9%)*
   C. It indicates the age of the rock (7.9%)
   D. It indicates the chemical composition of the rock (7.9%)

S3. The grain size of a clastic sedimentary rock is important because:
    (N = 104)

   **A. It reflects the environment in which the sediments were depos-
        ited (88.5%)**
   B. It is a function of the chemical composition of the rock (5.8%)
   C. It indicates the age of the rock (3.8%)
   D. It controls whether fossils can be found in the rock (1.9%)

S4. The grain size of a clastic sedimentary rock tells you: (N = 284)

   **A. The current energy where the sediments were deposited
        (60.2%)**
   B. *The chemical composition of the rock (20.4%)*
   C. *Whether or not the sediment formed from evaporation (11.6%)*
   D. How old the rock is (7.7%)

S5. Very poorly sorted sediments typically indicate which depositional environment? (N = 111)

    A. A glaciated area (75.7%)

    B. *A fast-flowing river (14.4%)*

    C. A tidal flat (5.4%)

    D. The deep ocean (4.5%)

S6. Shallow marine shelf sediments may include: (N = 106)

    A. Sand and mud, with burrows and trails (50.0%)

    B. *Very fine plankton remains and chert (27.4%)*

    C. *Poorly sorted sediments (18.9%)*

    D. Sand dunes and evaporites (3.8%)

S7. Which of the following is <u>NOT</u> a characteristic of limestone? (N = 111)

    A. Rarely formed at the surface of the Earth (45.9%)

    B. *Economically important (27.9%)*

    C. *Biogenic (17.1%)*

    D. Deposited in warm, shallow seas (9.0%)

S8. The presence of thick limestones in the stratigraphic record indicates to
geologists that in the past, that location was: (N = 429)

    A. A warm shallow marine setting (75.5%)

    B. *A slow-moving river (16.7%)*

    C. A fast-moving river (5.8%)

    D. A cold mountaintop (4.2%)

    E. A desert (3.3%)

    F. A sandy beach (2.6%)

Fossils 3 questions

F1. A key principle for interpreting both ancient rocks and fossils is uniformitarianism, also known as the uniformity of process, or actualism; this
principle states that: (N = 545)

    A. Present-day processes were also important in the past (50.6%)

    B. *The past is the key to predicting the future (40.2%)*

    C. Earth's natural laws have varied significantly over the past 4.5
billion years (6.1%)

    D. Ancient events were unique and unlike what happens today (3.1%)

F2. The "marine bias" in the fossil record indicates that marine organisms are much more likely to be fossilized than organisms that live on land. Why is this so? (N = 218)

    A. Marine organisms are more likely to be buried by sediment than life on land (84.4%)

    B. Bacterial decay cannot occur in saltwater (7.8%)

    C. Marine ecosystems experience very little deposition of sediment (5.5%)

    D. Marine organisms have hard parts while terrestrial organisms do not (2.3%)

F3. The "skeleton bias" in the fossil record indicates that: (N = 428)

    A. Organisms with hard parts are more likely to be preserved as fossils than soft-bodied ones (84.1%)

    B. Organisms with many parts to their skeletons are more likely to be fossilized (5.6%)

    C. People are more likely to collect fossils of vertebrates than invertebrate animals or plants (4.7%)

    D. Skeletons preserve more readily in saltwater conditions than in freshwater bodies (4.0%)

    E. Animals are more common than plants or single-celled organisms in the fossil record (3.3%)

Plate Tectonics 3 questions

PT1. Reversals in Earth's magnetic field: (N = 106)

    A. Can be used to date rocks (8.5%)

    B. Happen at irregular intervals (5.7%)

    C. Helped to demonstrate the phenomenon of seafloor spreading (3.8%)

    D. All of the above (82.1%)

PT2. Which of the following does **NOT** happen at a mid-ocean ridge? (N = 313)

    A. Old ocean crust is recycled back into the Earth (46.3%)

    B. *Two tectonic plates move away from each other (29.7%)*

    C. *Hot mantle rocks rise up toward the Earth's surface (16.0%)*

    D. New ocean crust is formed (8.0%)

PT3. A narrow zone of high, volcanic coastal mountains is likely to form at: (N = 333)

   A. An ocean–continent collision (69.1%)

   B. *A mid-ocean ridge (15.6%)*

   C. A continent–continent collision (9.9%)

   D. The center of a tectonic plate (5.4%)

Evolution and Natural Selection 6 questions

EV1. Evolution is: (N = 705)

   A. Descent with modification (48.9%)

   B. *The idea that life has continuously advanced and improved over the past 4 billion years (38.6%)*

   C. Natural selection, or "survival of the fittest" (8.5%)

   D. The idea that life originated on Earth by the spontaneous combination of molecules (4.0%)

EV2. Vestigial organs provide evidence of evolution because: (N = 394)

   A. They show that the ancestor of the organism must have looked different from how it does today (67.3%)

   B. *They are only found in fossils that are intermediate between two living groups, like reptiles and birds (16.0%)*

   C. *They demonstrate how poorly designed many organisms are (10.4%)*

   D. They change very rapidly, so we can observe the change directly in a lab (6.3%)

EV3. Which of the following is <u>NOT</u> a type of evidence for evolution? (N = 107)

   A. The extinction of the dinosaurs 65 Ma (86.9%)

   B. The embryological development of fish, chickens, and humans (7.5%)

   C. Bacterial resistance to antibiotics (3.7%)

   D. A fossil bird intermediate in anatomy between dinosaurs and living birds (1.9%)

EV4. What is the most important way that populations of animals acquire new variations upon which natural selection can act? (N = 455)

   A. Sexual recombination of genes from two parents in an offspring (50.3%)

B. *Pressure from the environment to change in order to survive (25.5%)*
C. *Mutations of genes (19.8%)*
D. Immigration of new individuals into a region (4.4%)

EV5. Which of the following statements about natural selection is <u>true</u>?
(N = 705)

A. *Natural selection causes individuals to develop traits that better suit their environment (43.1%)*

B. Natural selection can only produce a population better suited to its local, changing environment (29.6%)

C. *Natural selection always weeds out the weak and permits the strong to survive (15.3%)*

D. *Natural selection is an entirely random process (11.9%)*

EV6. It's called the <u>theory</u> of natural selection because: (N = 302)

A. It offers an explanation for how and why evolution happens (87.1%)

B. Scientists are still trying to prove that natural selection happens (5.3%)

C. Natural selection is a quantifiable law of nature (4.6%)

D. Natural selection is a synonym for evolution – the words are interchangeable (3.0%)

Precambrian 8 questions

PC1. How old is the Earth, according to radiometric dating of meteorites?
(N = 195)

A. 4.6 Ga (67.2%)
B. *3.5 Ga (24.8%)*
C. *4.1 Ga (13.3%)*
D. *4.0 Ga (10.5%)*
E. 6.0 Ga (5.6%)
F. 2.5 Ga (4.4%)

PC2. What geological objects do we radiometrically date to determine the age of the Earth? (N = 151)

A. Meteorites (61.6%)
B. *Stromatolites (13.9%)*
C. *Fossil bacteria from midocean ridges (12.6%)*
D. *Metamorphosed rocks from central Canada (11.9%)*

PC3. The origin of Earth's early atmosphere of water vapor, methane, ammonia, and carbon dioxide was: (N = 458)

    A. Volcanoes expelling gases from Earth's interior (65.5%)

    B. *Gasses accreting onto Earth as it formed from the solar nebula (22.3%)*

    C. Photosynthesis by bacteria (6.8%)

    D. The impact of a Mars-sized object with Earth early in its history (5.5%)

PC4. The buildup of atmospheric oxygen during the Precambrian is primarily attributed to: (N = 239)

    A. *Volcanic outgassing of the Earth's interior (39.3%)*

    B. Photosynthesis (36.4%)

    C. *Dissociation of water to form oxygen and hydrogen (13.8%)*

    D. *Weathering of carbonate-rich rocks (10.5%)*

PC5. Stromatolites are: (N = 180)

    A. Trace fossils of bacteria (43.3%)

    B. *Body fossils of bacteria (27.8%)*

    C. *The oldest rocks on Earth (18.3%)*

    D. *Iron-rich sedimentary rocks (10.6%)*

PC6. The endosymbiotic theory for the origin of eukaryotes argues that eukaryotic cells arose when: (N = 278)

    A. One bacterium tried to eat another one but couldn't digest it (64.4%)

    B. *Chemical conditions in the ocean favored larger, more complex cells (17.6%)*

    C. *Several bacteria chose to live together in order to form a multicellular life form (14.0%)*

    D. Bacteria began living inside the crust of the Earth (8.6%)

    E. Bacterial cells arose; they are the same age (3.3%)

PC7. According to the Snowball Earth hypothesis, what caused the global ice ages of the late Precambrian to end? (N = 202)

    A. Buildup of CO_2 from volcanoes led to an extreme greenhouse effect (75.2%)

B. *Breakup of the supercontinent Rodinia led to more burial of carbon in marine sediments (13.9%)*

C. The ice sheets absorbed more and more sunlight, eventually melting them. (8.9%)

D. Formation of extensive limestones warmed the world's oceans (2.0%)

PC8. Ediacarans were: (N = 104)

A. The first multicellular animals (69.2%)

B. *The first vertebrates (15.4%)*

C. *The first bacteria (12.5%)*

D. The first algae (2.9%)

Paleozoic 8 questions

PZ1. The Cambrian Explosion was: (N = 432)

A. The rapid appearance of modern animal groups in the fossil record (71.8%)

B. *The rapid increase of oxygen in the atmosphere (13.4%)*

C. *The first appearance of animals on land (10.2%)*

D. A series of large volcanic eruptions (4.6%)

PZ2. The Middle Cambrian Burgess Shale fossils are important because: (N = 148)

A. They show a greater range of animal forms than exists today (51.4%)

B. *They represent only the early ancestors of modern animal groups (29.6%)*

C. *They include the first animals that lived on land (22.3%)*

D. *They show that early animals were simple, primitive forms (16.0%)*

E. They show there were no predators in the Paleozoic Era (1.4%)

PZ3. An epifaunal suspension feeder: (N = 90)

A. Attaches to the seafloor and filters food from the water (55.6%)

B. *Floats in the water and grabs passing prey (25.6%)*

C. *Burrows into the sediment and filters food from the water (11.1%)*

D. Crawls on the seafloor and eats sediment (7.8%)

PZ4. Most members of Sepkoski's Paleozoic Marine Fauna (<u>post</u>-Cambrian animals) made their living by: (N = 98)

A. *Crawling on the seafloor and eating sediment (36.7%)*

B. Attaching to the seafloor and filtering food from the water (21.4%)

C. *Floating in the water and grabbing passing prey (21.4%)*

D. *Burrowing into the sediment and filtering food from the water (20.4%)*

PZ5. The following are all prominent members of the Paleozoic marine animal fauna **EXCEPT**: (N = 154)

    A. Sea urchins (69.5%)

    B. *Crinoids (11.7%)*

    C. Bryozoans (9.7%)

    D. Brachiopods (9.1%)

PZ6. The crash in atmospheric CO_2 levels during the Devonian was caused by: (N = 91)

    A. *More photosynthesis as land plants spread out for the first time (46.2%)*

    B. Increased chemical weathering of continents as land plants spread out for the first time (38.5%)

    C. An increase in oxygen levels as land vertebrates evolved (9.9%)

    D. A decrease in oxygen levels as land vertebrates evolved (5.5%)

PZ7. Vertebrate jawbones are structures modified from: (N = 194)

    A. Gill support bones (85.0%)

    B. Skull bones (7.7%)

    C. Hardened skin plates (6.7%)

    D. Non-vertebrate shells (4.4%)

    E. Fish scales (2.9%)

    F. Vertebrae (1.1%)

PZ8. Terrestrial vertebrates evolved from: (N = 91)

    A. Lobefin fish (71.4%)

    B. *Rayfin fish (19.8%)*

    C. Jawless fish (8.8%)

    D. 30-foot predatory fish (0.0%)

Mesozoic 11 questions

MZ1. Endothermy is related to which of the following characteristics?: (N = 217)

A. *Allowing body temperature to fluctuate with the environment (77.9%)*

B. *Giving birth to live young (11.5%)*

C. Processing food efficiently in the mouth (7.4%)

D. *Having seeds instead of spores (2.4%)*

E. *Laying eggs on dry land instead of in the water (3.4%)*

MZ2. Evidence for endothermy in mammal-like reptiles includes all of the following **EXCEPT**: (N = 318)

A. Ribs on most vertebrae (34.3%)

B. *Whisker pits on snout bones (30.8%)*

C. *Secondary palate (19.8%)*

D. *Limbs rotated under the body (15.1%)*

MZ3. Reptile and mammal jaw joints involve different bones. What type of jaw joint did transitional mammal-like reptiles have? (N = 41)

A. Both types of jaw joint (48.8%)

B. *The mammalian jaw joint (29.3%)*

C. *Their own unique third type of jaw joint (14.6%)*

D. The reptilian jaw joint (7.3%)

MZ4. The first, early dinosaurs: (N = 54)

A. Were small, bipedal forms (51.9%)

B. *Included some of the largest land animals that ever lived (18.5%)*

C. *Evolved from mammal-like reptiles (16.7%)*

D. *Arose long before mammals did (13.0%)*

MZ5. Evidence that at least <u>some</u> dinosaurs were warm-blooded includes all of the following **EXCEPT**: (N = 41)

A. *Parental care of offspring (39.0%)*

B. Had live birth instead of laying eggs (26.8%)

C. *Anatomical evidence of active running (22.0%)*

D. *Insulating feathers (12.2%)*

MZ6. Recent fossil finds in China show that some theropod dinosaurs closely resembled birds in having: (N = 91)

A. *Feathers (11.0%)*

B. Small body size (5.5%)

    C. Ability to glide through the air (1.1%)

    D. All of the above (82.4%)

MZ7. Feathers most likely first evolved: (N = 173)

    A. In dinosaurs for display or camouflage (49.7%)

    B. *In mammal-like reptiles to keep them warm (26.0%)*

    C. In Archaeopteryx *to help with gliding from tree to tree (16.8%)*

    D. In Mesozoic birds as they learned to fly (7.5%)

MZ8. *Archaeopteryx*, the earliest fossil bird, resembled a dinosaur in all the following **EXCEPT**: (N = 271)

    A. Was as large as a human (57.2%)

    B. *Long tail (17.0%)*

    C. *Claws on its hands (13.3%)*

    D. *Sharp teeth (12.5%)*

MZ9. Victims of the Cretaceous-Paleogene mass extinction include all of the following **EXCEPT**: (N = 324)

    A. Brachiopods (62.7%)

    B. *Ammonoids (20.4%)*

    C. *Pterosaurs (10.8%)*

    D. Dinosaurs (6.2%)

MZ10. Evidence for an asteroid impact at the end of the Cretaceous Period includes: (N = 342)

    A. Increased iridium concentrations (2.9%)

    B. The Chicxulub crater (1.5%)

    C. Shocked quartz grains (0.3%)

    D. All of the above (95.3%)

MZ11. According to the "Different Rules" model of mass extinctions, mammals survived the Cretaceous-Paleogene extinction because: (N = 350)

    A. They were small-bodied (60.0%)

    B. *They were warm-blooded endotherms, unlike dinosaurs (23.4%)*

    C. *They were evolutionary "newcomers" undergoing their initial radiation (16.6%)*

    D. *They were already in the process of evolving into larger, more diverse forms (16.0%)*

    E. They were more intelligent than dinosaurs (2.0%)

Cenozoic 6 questions

CZ1. An **increase** in the $\delta^{18}O$ value of fossil shells over time indicates: (N = 285)

    A. A decrease in water temperature (43.5%)

    B. *An increase in burial of organic carbon (22.6%)*

    C. *A decrease in burial of organic carbon (22.4%)*

    D. *A decrease in global ice volume (18.6%)*

    E. *A decrease in evaporation rate (15.4%)*

CZ2. A **decrease** in the $\delta^{18}O$ value of fossil shells over time indicates: (N = 83)

    A. *An increase in evaporation rate (32.5%)*

    B. *An increase in burial of organic carbon (25.3%)*

    C. An increase in water temperature (22.9%)

    D. *An increase in global ice volume (19.3%)*

CZ3. The initial pulse of Cenozoic global cooling that occurred in the Late Eocene (34 Ma) was probably related to: (N = 279)

    A. The rifting of Australia away from Antarctica (60.6%)

    B. *The formation of the Isthmus of Panama (24.7%)*

    C. The collision of India and Asia (7.9%)

    D. The formation of the Alps (6.8%)

CZ4. Cooling of Earth's climate in the Pliocene (2.5–3 Ma) was probably related to: (N = 98)

    A. The formation of the Isthmus of Panama (65.3%)

    B. *The collision of India and Asia (13.3%)*

    C. *The rifting of Australia away from Antarctica (13.3%)*

    D. The formation of the Alps (8.2%)

CZ5. The disappearance of South America's unique mammal fauna is related to: (N = 94)

    A. The migration of North American mammals across the Isthmus of Panama (75.5%)

B. *The formation of the volcanic Andes Mountains (11.7%)*

C. The superiority of marsupial over placental mammals (8.5%)

D. Competition with the giant phorusrachid ground birds (4.3%)

CZ6. Over the past few thousand years, we have been experiencing the climate conditions typical of: (N = 192)

A. An interglacial interval (50%)

B. *A glacial interval (27.6%)*

C. *One of the warmest times in Earth history (21.4%)*

D. One of the wettest times in Earth history (1.0%)

Appendix B
Evolution and Natural Selection Survey Results

The following 14 items were used in a pre-topic/post-topic survey pair to assess students' knowledge about evolution and natural selection in the introductory course "Life Through Time." Data shown here pool student responses over six semesters (fall 2012, spring 2013, fall 2013, fall 2014, fall 2015, and fall 2016).

Students were asked to individually and anonymously complete the survey before the class started discussing evolution, about one-third of the way through the semester (the "pre-topic" survey). Students then completed the same survey, again anonymously, in the last week of the semester, after they had learned about evolution, natural selection, and specific evolutionary events in the history of life on Earth (the "post-topic survey").

Students were informed that the survey was not graded and intended to be anonymous, so that they should feel free to react to each statement as honestly as possible. For each statement, students indicated whether they agreed or disagreed with it using a five-point Likert scale: 1-Strongly Agree, 2-Tend to Agree, 3-Don't Know, 4-Tend to Disagree, 5-Strongly Disagree.

The survey items are modified from a resource on misconceptions about evolution within the Understanding Evolution website produced by the University of California Museum of Paleontology and available at: http://evolution.berkeley.edu/evolibrary/misconceptions_faq.php. All 14 statements are, in fact, common misconceptions about evolution and natural selection.

Percentages of students agreeing or disagreeing with each item are provided in Tables B1 (pre-topic) and B2 (post-topic). Of particular note are items that appear bimodal (with similar percentages of students on the "agree" and "disagree" sides, indicated in **boldface** in Table B1); these may be especially polarizing concepts in class. One should also consider those items with a large percentage of students indicating they don't know (items with more than 33% "don't know" in *italic* in Table B1). Note that the percentage of students marking "don't know" drops for every item except item 14 in the post-topic survey.

Table B1 Percentage of students giving each response on the pre-topic survey

Pre-Topic Survey	1-Strongly Agree	2-Tend to Agree	3-Don't Know	4-Tend to Disagree	5-Strongly Disagree	N
1. Evolution is a theory about the origin of life.	38	36	7	12	7	616
2. Evolution is like a climb up a ladder of progress; organisms are always getting better.	18	43	15	18	6	616
3. Evolution means that life changed by chance.	4	17	22	37	21	616
4. Natural selection involves organisms trying to adapt.	31	45	10	8	5	614
5. Natural selection gives organisms what they need.	11	36	28	17	9	614
6. Evolution is just a theory.	**16**	**25**	**18**	**25**	**15**	615
7. *Most biologists have rejected Darwinism (i.e., no longer really agree with the ideas put forth by Darwin).*	*3*	*6*	*43*	*31*	*17*	615
8. Gaps in the fossil record disprove evolution.	1	5	27	33	35	615
9. The theory of evolution is flawed, but scientists won't admit it.	3	13	29	29	26	613
10. Evolution is not science because it is not observable or testable.	1	7	11	38	43	616
11. Evolution leads to immoral behavior. If children are taught that they are animals, they will behave like animals.	1	4	8	24	63	614
12. *Evolution supports the idea that might makes right and rationalizes the oppression of some people by others.*	*2*	*10*	*40*	*18*	*30*	612
13. Evolution and religion are incompatible.	11	17	19	31	23	605
14. Teachers should teach both evolution and creationism and let students decide for themselves.	**21**	**24**	**15**	**15**	**25**	604

Table B2 Percentage of students giving each response on the post-topic survey

Post-Topic Survey	1-Strongly Agree	2-Tend to Agree	3-Don't Know	4-Tend to Disagree	5-Strongly Disagree	N
1. Evolution is a theory about the origin of life.	27	31	5	18	20	520
2. Evolution is like a climb up a ladder of progress; organisms are always getting better.	6	22	5	31	35	524
3. Evolution means that life changed by chance.	8	20	12	33	25	522
4. Natural selection involves organisms trying to adapt.	19	36	6	21	19	521
5. Natural selection gives organisms what they need.	5	20	10	38	27	518
6. Evolution is just a theory.	12	15	10	32	31	522
7. Most biologists have rejected Darwinism (i.e., no longer really agree with the ideas put forth by Darwin).	5	13	21	35	25	520
8. Gaps in the fossil record disprove evolution.	1	6	9	38	46	525
9. The theory of evolution is flawed, but scientists won't admit it.	3	11	12	35	38	524
10. Evolution is not science because it is not observable or testable.	2	3	6	30	59	525
11. Evolution leads to immoral behavior. If children are taught that they are animals, they will behave like animals.	4	5	6	19	66	524
12. Evolution supports the idea that might makes right and rationalizes the oppression of some people by others.	2	12	27	16	43	523
13. Evolution and religion are incompatible.	11	14	14	30	31	525
14. Teachers should teach both evolution and creationism and let students decide for themselves.	26	22	16	13	23	524

Appendix C
Student-Reported Misconceptions

On the last day of class in the introductory course "The Geologic History of Dinosaurs," students were asked to complete a final "Course Reflection" questionnaire. One question asked students to reflect on a misconception they had before they took the course that changed as a result of the course. Students were informed that the survey was not graded and intended to be anonymous, and that they should feel free to react to each statement as honestly as possible.

Student responses were read and categorized into themes, listed later in this appendix. Data shown here pool 238 student responses collected over four semesters (spring 2014, spring 2015, spring 2016, and spring 2017).

Student Prompt: Describe one **misconception** you had (about dinosaurs, paleontology, geology, evolution, etc.) in January that has changed because of this course.

Nature of Science

1) Science is mostly about collecting facts; most of what scientists do is collect facts and make observations.
2) Hypotheses are just educated guesses.
3) Science is only interesting to scientists.

Geology

4) Sedimentary rocks can be directly dated to get an age in millions of years ago.
5) All radiometric dating is carbon dating; carbon dating can date any material of any age.
6) Pangea existed throughout the Mesozoic, dinosaur time.

Paleontology

7) Every paleontologist studies dinosaurs; every paleontologist studies all ancient life.
8) Fossils are extremely rare.
9) Fossil specimens are mostly complete skeletons.
10) Only trained paleontologists are allowed to look for fossils.
11) Paleontologists only look for fossils and dig them up.
12) Paleontology is just guesswork and speculation because of the lack of evidence, inability to observe directly.

13) The only things you can learn from a fossil are what kind of animal it was and when it lived.
14) We already know everything there is to know about ancient life; paleontologists do not need to revise their understanding of the past.

Dinosaurs

15) Everything living during the time of the dinosaurs (e.g., pterosaurs, marine reptiles) was a dinosaur; all ancient organisms are called dinosaurs.
16) Very few different kinds of dinosaurs existed, one of each "type" (e.g., one sauropod, one duckbill).
17) Most or all dinosaurs were meat-eaters.
18) All dinosaurs lived at the same time and became extinct at the same time.
19) Dinosaurs were never very successful and did not exist for very long.
20) Dinosaurs dominated life for the entire Mesozoic era, or for all of Earth history up until their extinction.
21) All dinosaurs were large-bodied, cold-blooded, scaly, and not intelligent.
22) All dinosaurs lived in tropical forests.
23) The dinosaurs in the film *Jurassic Park* are accurately portrayed; raptors were very smart and *T. rex* could run as fast as a car.
24) Birds and dinosaurs are not related; birds are not dinosaurs; dinosaurs are more closely related to modern reptiles than to modern birds.
25) Birds evolved from pterosaurs.
26) Flight evolved in one step, without any intermediate stages like gliding.
27) All paleontologists agree that an asteroid impact caused the extinction of the dinosaurs.
28) Mammals evolved after dinosaurs became extinct.
29) Scientists will soon be able to clone dinosaurs, like in *Jurassic Park*.

Evolution

30) Organisms intentionally select which traits to evolve.
31) The environment causes evolution to happen.
32) Evolutionary change happens quickly, from one generation to the next.
33) Evolution and natural selection are synonyms.
34) Dinosaurs are so different from each other (e.g., plant-eaters vs. meat-eaters) that they could not possibly all have a common ancestor.

References

Anderson, D. L., Fisher, K. M., and Norman, G. L. (2002). Development and validation of the conceptual inventory of natural selection. *Journal of Research in Science Teaching*, **39**, 952–978.

Anderson, S. W., and Libarkin, J. C. (2016). Conceptual mobility and entrenchment in introductory geoscience courses: New questions regarding physics' and chemistry's role in learning Earth science concepts. *Journal of Geoscience Education*, **64**, 74–86.

Arthurs, L., Hsia, J. F., and Schweinle, W. (2015). The Oceanography Concept Inventory: A semicustomizable assessment for measuring student understanding of oceanography. *Journal of Geoscience Education*, **63**, 310–322.

Baldwin, K. A., and Cooper, C. M. (2014). Online and on-campus historical geology students' prior ideas about global climate change. *Journal of Geoscience Education*, **62**, 410–416.

Bilici, S. C., Armagan, F. O., Cakir, N. K., and Yuruk, N. (2011). The development of an Astronomy Concept Inventory (ACI). *Procedia Social and Behavioral Sciences*, **15**, 2454–2458.

Bodzin, A. M., Anastasio, D., Sahagian, D., Peffer, T., Dempsey, C., and Steelman, R. (2014). Investigating climate change understandings of urban middle-level students. *Journal of Geoscience Education*, **62**, 417–430.

Bransford, J., Brown, A. L., and Cocking, R. R. (eds.). (2000). *How People Learn: Brain, Mind, Experience, and School*, Washington, DC: National Research Council.

Capps, D. K., McAllister, M., and Boone, W. J. (2013). Alternative conceptions concerning the Earth's interior exhibited by Honduran students. *Journal of Geoscience Education*, **61**, 231–239.

Cheek, K. A. (2010). Commentary: A summary and analysis of twenty-seven years of geoscience conceptions research. *Journal of Geoscience Education*, **58**, 122–134.

Chi, M. T. H., Slotta, J. D., and de Leeuw, N. (1994). From things to processes: A theory of conceptual change for learning science concepts. *Learning and Instruction*, **4**, 27–43.

Clark, S. K., Libarkin, J. C., Kortz, K. M., and Jordan, S. C. (2011). Alternative conceptions of plate tectonics held by nonscience undergraduates. *Journal of Geoscience Education*, **59**, 251–262.

Climate Literacy Network. (2009). *Climate Literacy: The Essential Principles of Climate Sciences,* http://oceanservice.noaa.gov/education/literacy/clima te_literacy.pdf. Accessed December 1, 2017.

Dahl, J., Anderson, S. W., and Libarkin, J. C. (2005). Digging into Earth science: Alternative conceptions held by K-12 teachers. *Journal of Science Education,* **6,** 65–68.

Dahl, R. M. (2018). Education research as applied to paleontology education: How people learn and how we can teach more effectively. Paleontological Society Short Course: Pedagogy and Technology in the Modern Paleontology Classroom. *Elements of Paleontology.*

D'Avanzo, C. (2008). Biology concept inventories: Overview, status, and next steps. *BioScience,* **58,** 1079–1085.

Dodick, J. T., and Orion, N. (2003). Measuring student understanding of "deep time." *Science Education,* **87,** 708–731.

Dole, J. A., and Sinatra, G. M. (1998). Reconceptualizing change in the cognitive construction of knowledge. *Educational Psychologist,* **33,** 109–128.

Donovan, M. S., and Bransford, J. D. (2005). *How Students Learn Science in the Classroom,* Washington, DC: National Academies Press.

Dove, J. E. (1998). Students' alternative conceptions in Earth science: A review of research and implications for teaching and learning. *Research Papers in Education,* **13,** 183–201.

Driver, R., and Easley, J. (1978). Pupils and paradigms: A review of literature related to concept development in adolescent science students. *Studies in Science Education,* **5,** 61–84.

Driver, R., and Erickson, G. (1983). Theories-in-action: Some theoretical and empirical issues in the study of students' conceptual frameworks in science. *Studies in Science Education,* **10,** 37–60.

Driver, R., Guesne, E., and Tiberghien, A. (1985). *Children's Ideas in Science,* Milton Keynes, UK: Open University Press.

Driver, R., Squires, A., Rushworth, P., and Wood-Robinson, V. (1994). *Making Sense of Secondary Science: Research into Children's Ideas,* London: Routledge.

Duit, R., and Treagust, D. F. (2003). Conceptual change: A powerful frame-work for improving science teaching and learning. *International Journal of Science Education,* **25,** 671–688.

Earth Science Literacy Initiative. (2010). *Earth Science Literacy Principles: The Big Ideas and Supporting Concepts of Earth Science,* www.earthscien celiteracy.org. Accessed November 28, 2017.

Felzmann, D. (2017). Students' conceptions of glaciers and ice ages: Applying the Model of Educational Reconstruction to improve learning. *Journal of Geoscience Education*, **65**, 322–355.

Flammer, L. (1999). *Science Knowledge Survey*, www.indiana.edu/~ensiweb/lessons/sci.tst.html. Accessed November 30, 2017.

Francek, M. (2013). A compilation and review of over 500 geoscience misconceptions. *International Journal of Science Education*, **35**, 31–64.

Fulwiler, T. (1987). *Teaching with Writing*, Portsmouth, NH: Heineman.

Garvin-Doxas, K., and Klymkowsky, M. W. (2008). Understanding randomness and its impact on student learning: Lessons learned from building the Biology Concept Inventory (BCI). *CBE – Life Sciences Education*, **7**, 227–233.

Geoscience Concept Inventory Wiki. (No Date). https://geoscienceconceptin ventory.wikispaces.com/. Accessed November 30, 2017.

Hestenes, D., Wells, M., and Swackhammer, G. (1992). Force concept inventory. *The Physics Teacher*, **30**, 141–151.

Hewson, P. W. (1981). A conceptual change approach to learning science. *European Journal of Science Education*, **3**, 383–396.

Hewson, P. W. (1992). Conceptual change in science teaching and teacher education. Paper presented in June 1992, National Center for Educational Research, Documentation, and Assessment, Madrid, Spain, www.learner .org/workshops/lala2/support/hewson.pdf. Accessed November 27, 2017.

Hewson, P. W., and Hewson, M. G. A'B. (1988). An appropriate conception of teaching science: A view from studies of science learning. *Science Education*, **72**, 597–614.

Hidalgo, A. J., and Otero, J. (2004). An analysis of the understanding of geological time by students at secondary and post-secondary level. *International Journal of Science Education*, **26**, 845–857.

Johnson, J. K., and Reynolds, S. J. (2005). Concept sketches: Using student- and instructor-generated, annotated sketches for learning, teaching, and assessment in geology courses. *Journal of Geoscience Education*, **53**, 85–95.

Jolley, A., Jones, F., and Harris, S. (2013). Measuring student knowledge of landscapes and their formation timespans. *Journal of Geoscience Education*, **61**, 240–251.

Keeley, P. (2005). *Science Curriculum Topic Study: Bridging the Gap between Standards and Practice*, Thousand Oaks, CA: Corwin Press.

Keeley, P. (2008). *Science Formative Assessment: 75 Practical Strategies for Linking Assessment, Instruction, and Learning*, Thousand Oaks, CA: Corwin Press.

Keeley, P. (2011). *Uncovering Student Ideas*, www.uncoveringstudentideas .org/. Accessed November 30, 2017.

Keeley, P. (2015a). Mountaintop fossil: A puzzling phenomenon. *Science and Children*, **53**(4), 24–26.

Keeley, P. (2015b). *Science Formative Assessment, Vol. 2: 50 More Strategies for Linking Assessment, Instruction, and Learning*, Thousand Oaks, CA: Corwin Press.

Keeley, P., Eberle, F., and Dorsey, C. (2008). *Uncovering Student Ideas in Science, Vol. 3: Another 25 Formative Assessment Probes*, Arlington, VA: NSTA Press.

Keeley, P., Eberle, F., and Farrin, L. (2005). *Uncovering Student Ideas in Science: 25 Formative Assessment Probes*, Arlington, VA: NSTA Press.

Keeley, P., Eberle, F., and Tugel, J. (2007). *Uncovering Student Ideas in Science, Vol. 2: 25 More Formative Assessment Probes*, Arlington, VA: NSTA Press.

Keeley, P., and Tugel, J. (2009). *Uncovering Student Ideas in Science, Vol. 4: 25 New Formative Assessment Probes*, Arlington, VA: NSTA Press.

Kuh, G. D. (2008). *High-Impact Educational Practices: What They Are, Who Has Access to Them, and Why They Matter*, Washington, DC: Association of American Colleges and Universities.

Lambert, J. L., Lindgren, J., and Bleicher, R. (2012). Assessing elementary science methods students' understanding about global climate change. *International Journal of Science Education*, **34**, 1167–1187.

Lee, O. (2001). Preface: Culture and language in science education: What do we know and what do we need to know? *Journal of Research in Science Teaching*, **38**, 499–501.

Lee, O., Maerten-Rivera, J., Buxton, C., Penfield, R., and Secada, W. G. (2009). Urban elementary teachers' perspectives on teaching science to English language learners. *Journal of Science Teacher Education*, **20**, 263–286.

Libarkin, J. C. (2006). College student conceptions of geological phenomena and their importance in classroom instruction. *Planet*, **17**(1), 6–9.

Libarkin, J. (2008). *Concept Inventories in Higher Education Science*. National Research Council, www7.nationalacademies.org/bose/ Libarkin_CommissionedPaper.pdf. Accessed November 28, 2017.

Libarkin, J. C., and Anderson, S. W. (2005). Assessment of learning in entry-level geoscience courses: Results from the Geoscience Concept Inventory. *Journal of Geoscience Education*, **53**, 394–401.

Libarkin, J. C., and Anderson, S. W. (2007a). Development of the Geoscience Concept Inventory. *Proceedings of the National STEM Assessment Conference*, Washington DC, October 19–21, 2006, 148–158.

Libarkin, J. C., and Anderson, S. W. (2007b). The Geoscience Concept Inventory: Application of Rasch Analysis to concept inventory development in higher education. In X. Liu and W. Boone, eds., *Applications of Rasch Measurement in Science Education*, Maple Grove, MN: JAM Publishers, pp. 45–73.

Libarkin, J. C., Anderson, S. W., Science, J. D., Beilfuss, M., and Boone, W. (2005). Qualitative analysis of college students' ideas about the Earth: Interviews and open-ended questionnaires. *Journal of Geoscience Education*, **53**, 17–26.

Libarkin, J., Jardeleza, S. E., and McElhinny, T. L. (2014). The role of concept inventories in course assessment. In V. C. H. Tong, ed., *Geoscience Research and Education: Teaching at Universities*, Innovations in Science Education and Technology, vol. 20, Dordrecht: Springer, pp. 275–297.

Libarkin, J. C., and Kurdziel, J. P. (2001). Research methodologies in science education: Assessing students' alternative conceptions. *Journal of Geoscience Education*, **49**, 378–383.

Libarkin, J. C., and Kurdziel, J. P. (2002). Research methodologies in science education: The qualitative-quantitative debate. *Journal of Geoscience Education*, **50**, 78–86.

Libarkin, J. C., and Kurdziel, J. P. (2006). Ontology and the teaching of Earth system science. *Journal of Geoscience Education*, **54**, 408–413.

Libarkin, J. C., Kurdziel, J. P., and Anderson, S. W. (2007). College student conceptions of geological time and the disconnect between ordering and scale. *Journal of Geoscience Education*, **55**, 413–422.

Libarkin, J. C., Ward, E. M. G., Anderson, S. W., Kortemeyer, G., and Raeburn S. P. (2011). Revisiting the Geoscience Concept Inventory: A call to the community. *GSA Today*, **21**(8), 26–28.

Lindell, R., Peak, E., and Foster, T. (2007). Are they all created equal? A comparison of different concept inventory development methodologies. *Physics Education Research Conference, American Institute of Physics*, **883**, 14–17.

Luykx, A., Lee, O., and Edwards, U. (2008). Lost in translation: Negotiating meaning in a beginning ESOL science classroom. *Educational Policy*, **22**, 640–674.

Martínez, P., Bannan, B., and Kitsantas, A. (2012). Bilingual students' ideas and conceptual change about slow geomorphological changes caused by water. *Journal of Geoscience Education*, **60**, 54–66.

Mazur, E. (1997). *Peer Instruction: A User's Manual*, Series in Educational Innovation, Upper Saddle River, NJ: Prentice Hall.

McConnell, D. A., Steer, D. N., Owens, K. D., and Knight, C. C. (2005). How students think: Implications for learning in introductory geoscience courses. *Journal of Geoscience Education*, **53**, 462–470.

McConnell, D. A., Steer, D. N., Owens, K. D., Knott, J. R., Van Horn, S., Borowski, W., Dick, J., Foos, A., Malone, M., McGrew, H., Greer, L., and Heaney, P. J. (2006). Using conceptests to assess and improve student conceptual understanding in introductory geoscience courses. *Journal of Geoscience Education*, **54**, 61–68.

McCuin, J. L., Hayhoe, K., and Hayhoe, D. (2014). Comparing the effects of traditional vs. misconceptions-based instruction on student understanding of the greenhouse effect. *Journal of Geoscience Education*, **62**, 445–459.

McDermott, M. (2010). More than writing-to-learn. *The Science Teacher*, **77**(1), 32–36.

McNeal, K. S., Spry, J. M., Mitra, R., and Tipton, J. L. (2014). Measuring student engagement, knowledge, and perceptions of climate change in an introductory environmental geology course. *Journal of Geoscience Education*, **62**, 655–667.

Mulford, D. R., and Robinson, W. R. (2002). An inventory for misconceptions in first-semester general chemistry. *Journal of Chemical Education*, **76**, 739–744.

Nathan, M. J., and Alibali, M. W. (2010). Learning sciences. *Wiley Interdisciplinary Reviews: Cognitive Science*, **1**, 329–345.

National Survey of Student Engagement. (2005). *Exploring Different Dimensions of Student Engagement*, Bloomington, IN: Indiana University Center for Postsecondary Research.

Nelson-Laird, T. F., Shoup, R., Kuh, G. D., and Schwartz, M. J. (2008). The effects of discipline on deep approaches to student learning and college outcomes. *Research in Higher Education*, **49**, 469–494.

Pavelich, M., Jenkins, B., Birk, J., Bauer, R., and Krause, S. (2004). Development of a chemistry concept inventory for use in chemistry, materials and other engineering courses. *Proceedings of the American Society for Engineering Education Annual Conference and Exposition*, Paper #2004–1907.

Perez, K. E., Hiatt, A., David, G. K., Trujillo, C., French, D. P., Terry, M., and Prince, R. M. (2013). The EvoDevoCI: A concept inventory for gauging students' understanding of evolutionary developmental biology. *CBE – Life Sciences Education*, **12**, 665–675.

Petcovic, H. L., and Ruhf, R. J. (2008). Geoscience conceptual knowledge of preservice elementary teachers: Results from the Geoscience Concept Inventory. *Journal of Geoscience Education*, **56**, 251–260.

Phillips, W. (1991). Earth science misconceptions. *The Science Teacher*, **58**(2), 21–23.

Piaget, J. (1967). *Logique et Connaissance Scientifique, Encyclopédie de la Pléiade*, Paris: Éditions Gallimard.

Piaget, J. (1973). *To Understand Is to Invent*, New York, NY: Grossman.

Piaget, J., and Inhelder, B. (1969). *The Psychology of the Child*. Translated from the French by H. Weaver. New York, NY: Basic Books, Inc.

Posner, G. J., Strike, K. A., Hewson, P. W., and Gertzog, W. A. (1982). Accommodation of a scientific conception: Toward a theory of conceptual change. *Science Education*, **66**, 211–227.

Raia, F. (2005). Students' understanding of complex dynamic systems. *Journal of Geoscience Education*, **53**, 297–308.

Rebich, S., and Gautier, C. (2005). Concept mapping to reveal prior knowledge and conceptual change in a mock summit course on global climate change. *Journal of Geoscience Education*, **53**, 355–365.

Reichert, C., Cervato, C., Larsen, M., and Niederhauser, D. (2014). Conceptions of atmospheric carbon budgets: Undergraduate students' perceptions of mass balance. *Journal of Geoscience Education*, **62**, 460–468.

Reinfried, S., and Schuler, S. (2009). Die Ludwigsburg-Luzerner Bibliographie zur Alltagsvorstellungsforschung in den Geowissenschaften – ein Projekt zur Erfassung der internationalen Forschungsliteratur [The Ludwigsburg-Lucerne bibliography on conceptual change research in the geosciences – A project to establish a comprehensive collection of international research papers in the field]. *Geographie und ihre Didaktik*, **37**, 120–135, www.ph-ludwigsburg.de/llbg.html. Accessed November 28, 2017. [Note: In German. Updated listing of papers is available for download from this website in pdf, Word, or EndNote formats.]

Schoon, K. J. (1992). Students' alternative conceptions of earth and space. *Journal of Geological Education*, **40**, 209–214.

Sell, K. S., Herbert, B. E., Stuessy, C. L., and Schielack, J. (2006). Supporting student conceptual model development of complex Earth systems through the use of multiple representations and inquiry. *Journal of Geoscience Education*, **54**, 396–407.

Sexton, J. M. (2012). College students' conceptions of the role of rivers in canyon formation. *Journal of Geoscience Education*, **60**, 168–178.

Sibley, D. F. (2005). Visual abilities and misconceptions about plate tectonics. *Journal of Geoscience Education*, **53**, 471–477.

Smith, G. A., and Bermea, S. B. (2012). Using students' sketches to recognize alternative conceptions about plate tectonics persisting from prior instruction. *Journal of Geoscience Education*, **60**, 350–359.

Smith, J. I., and Tanner, K. (2010). The problem of revealing how students think: Concept inventories and beyond. *CBE – Life Sciences Education*, **9**, 1–5.

Smith, M. K., Wood, W. B., and Knight, J. K. (2008). The Genetics Concept Assessment: A new concept inventory for gauging student understanding of genetics. *CBE – Life Sciences Education*, **7**, 422–430.

Solano-Flores, G., and Nelson-Barber, S. (2001). On the cultural validity of science assessments. *Journal of Research in Science Teaching*, **38**, 553–573.

Steer, D. N., Knight, C. C., Owens, K. D., and McConnell, D. A. (2005). Challenging students [*sic*] ideas about Earth's interior structure using a model-based, conceptual change approach in a large class setting. *Journal of Geoscience Education*, **53**, 415–421.

Stepans, J. (2008). *Targeting Students' Physical Science Misconceptions Using the Conceptual Change Model*, 3rd edn, Saint Cloud, MN: Saiwood Publications.

Svinicki, M. (1995). Using cognitive theories to improve teaching. *The Teaching Professor*, **9**(4), 3–4.

Teed, R., and Slattery, W. (2011). Changes in geologic time understanding in a class for preservice teachers. *Journal of Geoscience Education*, **59**, 151–162.

Treagust, D. (1986). Evaluating students' misconceptions by means of diagnostic multiple choice items. *Research in Science Education*, **16**, 199–207.

Treagust, D., and Duit, R. (2008). Conceptual change: A discussion of theoretical, methodological and practical challenges for science education. *Cultural Studies of Science Education*, **3**, 297–328.

Trend, R. D. (1998). An investigation into understanding of geological time among 10- and 11-year-old children. *International Journal of Science Education*, **20**, 973–988.

Trend, R. D. (2000). Conceptions of geological time among primary teacher trainees, with reference to their engagement with geoscience, history, and science. *International Journal of Science Education*, **22**, 539–555.

Trend, R. D. (2001). Deep time frameworks: A preliminary study of U.K. primary teachers' conceptions of geological time and perceptions of geoscience. *Journal of Research in Science Teaching*, **38**, 191–221.

Undersander, M. A., Kettler, R. M., and Stains, M. (2017). Exploring the item order effect in a geoscience concept inventory. *Journal of Geoscience Education*, **65**, 292–303.

Understanding Evolution. (2017). *Misconceptions about Evolution*, https://evolution.berkeley.edu/evolibrary/misconceptions_teacherfaq.php. Accessed November 30, 2017.

Understanding Science. (2017). *Misconceptions about Science*, https://undsci .berkeley.edu/teaching/misconceptions.php. Accessed November 30, 2017.

Vosniadou, S., and Brewer, W. F. (1992). Mental models of the Earth: A study of conceptual change in childhood. *Cognitive Psychology*, **24**, 535–385.

Ward, E. M. G., Libarkin, J. C., Raeburn, S., and Kortemeyer, G. (2010). The Geoscience Concept Inventory Web Center provides new means for student assessment. *eLearningPapers*, 20, 1–14.

Weimer, M. (2002). *Learner-Centered Teaching: Five Key Changes to Practice*, San Francisco, CA: Jossey-Bass.

Wiggins, G., and McTighe, J. (2005). *Understanding by Design*, expanded 2nd edn, Alexandria, VA: Association for Supervision and Curriculum Development.

Wild, T. A., Hilson, M. P., and Farrand, K. M. (2013). Conceptual understanding of geological concepts by students with visual impairments. *Journal of Geoscience Education*, **61**, 222–230.

Yacobucci, M. M. (2012). Using active learning strategies to promote deep learning in the undergraduate paleontology classroom. In M. M. Yacobucci and R. Lockwood, eds., *Teaching Paleontology in the 21st Century*, Paleontological Society Special Publications, 12, 135–153.

Yin, Y., Shavelson, R. J., Ayala, C. C., Ruiz-Primo, M. A., Brandon, P. R., Furtak, E. M., Tomita, M. K., and Young, D. B. (2008). On the impact of formative assessment on student motivation, achievement, and conceptual change. *Applied Measurement in Education*, **21**, 335–359.

Zull, J. E. (2004). The art of changing the brain. *Educational Leadership*, **62**, 68–72.

Acknowledgments

I first want to thank all of my students. I have taught nearly 1,500 students in my introductory courses over the past seven years. They have provided me with invaluable insight into how people think about science, fossils, evolution, and the Earth. It is their work that provides the data presented in this Element. I am also grateful to two anonymous reviewers and A. Avruch, who provided useful feedback on the manuscript. Finally, I wish to thank the organizers of this short course for promoting new ways of approaching undergraduate paleontology education.

Cambridge Elements $\overline{\overline{}}$

Elements of Paleontology

Editor-in-Chief

Colin D. Sumrall
University of Tennessee

About the Series
The Elements of Paleontology series is a publishing collaboration between the Paleontological Society and Cambridge University Press. The series covers the full spectrum of topics in paleontology and paleobiology, and related topics in the Earth and life sciences of interest to students and researchers of paleontology.

The Paleontological Society is an international nonprofit organization devoted exclusively to the science of paleontology: invertebrate and vertebrate paleontology, micropaleontology, and paleobotany. The Society's mission is to advance the study of the fossil record through scientific research, education, and advocacy. Its vision is to be a leading global advocate for understanding life's history and evolution. The Society has several membership categories, including regular, amateur/avocational, student, and retired. Members, representing some 40 countries, include professional paleontologists, academicians, science editors, Earth science teachers, museum specialists, undergraduate and graduate students, postdoctoral scholars, and amateur/avocational paleontologists.

Paleontological
S O C I E T Y

Cambridge Elements $\equiv$

Elements of Paleontology

Elements in the Series

These Elements are contributions to the Paleontological Short Course on *Pedagogy and Technology in the Modern Paleontology Classroom* (organized by Phoebe A. Cohen, Rowan Lockwood, and Lisa Boush), convened at the Geological Society of America Annual Meeting in November 2018 (Indianapolis, Indiana USA).

Flipping the Paleontology Classroom: Benefits, Challenges, and Strategies
Matthew E. Clapham

Integrating Macrostrat and Rockd into Undergraduate Earth Science Teaching
Phoebe A. Cohen, Rowan Lockwood, and Shanan Peters

Student-Centered Teaching in Paleontology and Geoscience Classrooms
Robyn Mieko Dahl

Beyond Hands On: Incorporating Kinesthetic Learning in an Undergraduate Paleontology Class
David W. Goldsmith

Incorporating Research into Undergraduate Paleontology Courses: Or a Tale of 23,276 Mulinia
Patricia H. Kelley

Utilizing the Paleobiology Database to Provide Educational Opportunities for Undergraduates
Rowan Lockwood, Phoebe A. Cohen, Mark D. Uhen, and Katherine Ryker

Integrating Active Learning into Paleontology Classes
Alison N. Olcott

Dinosaurs: A Catalyst for Critical Thought
Darrin Pagnac

Confronting Prior Conceptions in Paleontology Courses
Margaret M. Yacobucci

The Neotoma Paleoecology Database: A Research Outreach Nexus
Simon J. Goring, Russell Graham, Shane Oeffler, Amy Myrbo, James S. Oliver, Carol Ormond, and John W. Williams

Equity, Culture, and Place in Teaching Paleontology: Student-Centered Pedagogy for Broadening Participation
Christy C. Visaggi

A full series listing is available at: www.cambridge.org/EPLY

For EU product safety concerns, contact us at Calle de José Abascal, 56–1°,
28003 Madrid, Spain or eugpsr@cambridge.org.